T0331083

Blending business and academics, Artificial Intelligence for Business Optimization: Research and Applications *fills the need for authoritative information in an emerging market. The book is a valuable contribution on the academic, business and ICT services. The authors have a reputation of thought leadership, with academic and business credibility.*

Keith Sherringham (BSc. Hons, FACS)
Senior Vice President at Citi
Greater Sydney Area, Australia

This book makes a unique contribution in the field of Artificial Intelligence (AI) by focusing on optimizing business processes. The business context in the discussions herein is excellent. The material presented here has practical applications in health, education, sustainability and many other such areas that are important for the quality of life on Earth.

Andy Lyman
Chairman of the Board for All Point POS
Florida, USA

The authors have done an excellent job in discussing the application of AI to business. Crucial topics such as leadership and business strategies in optimization are very well presented. Dynamicity in learning will open up many new areas of research.

Dr. Anurag Agarwal
Professor, Department of Information Systems &
Operations Management
Florida Gulf Coast University
Florida, USA

I am a believer in the power of Artificial Intelligence for business optimization from a strategic standpoint. This book is a journey from data to decisions. The unique business perspective shown by the authors is invaluable in understanding Artificial Intelligence in practical application in business organizations. Adopting data-driven culture and the value of leadership and change management in the context of AI make this book unique.

Jean Kabongo, PhD
Campus Dean and Professor, Muma College of Business,
Sarasota-Manatee
University of South Florida
Florida, USA

At USF, especially in my college, the focus is squarely on application of technology to business. This book by Drs. Unhelkar and Gonsalves, does a fine job of demonstrating the application of Artificial Intelligence to the challenges of Business Optimization. AI and ML have already transcended Automation and need to be increasingly applied in business process optimization - as discussed and demonstrated in this book. This book makes a valuable contribution for both practitioners and researchers.

Dr. Kaushik Dutta
Professor & Muma Fellow
Director - School of Information Systems & Management,
Muma College of Business
University of South Florida
Florida, USA

Artificial Intelligence for Business Optimization

Artificial Intelligence for Business Optimization
Research and Applications

Bhuvan Unhelkar
Tad Gonsalves

CRC Press
Taylor & Francis Group
Boca Raton London New York

CRC Press is an imprint of the
Taylor & Francis Group, an **informa** business

FAMILY

[BU]: Thanks to my family for their support and good wishes: Asha (wife), Sonki (daughter), Keshav (son), and Chinar (sister-in-law); and our dog Benji. This book is dedicated to my extended family!

[TG]: This book is dedicated to the new generation in our family: Savio, Glenda, Samson, Johnson, Qutandy, Janice, Lester, and Pearl. Thank you for your constant support and wishes.

DEDICATION

Trivikrama (TV), Jayalakshmi, and Shankar [BU]

Isabel and Paulo [TG]

Contents

Foreword by Andy Lyman

The book in your hand makes a unique contribution in the field of Artificial Intelligence (AI). The mention of the word AI is almost synonymous with imagery of data centers, complex analytics, neural networks, and automation. The age-old stories of robots taking over humans and computers beating chess masters add to the spice. I am fortunate to be witnessing the growth of AI as a technology as it coincides with my own growth from that of technology expertise to managing and growing a business. Businesses, for most part, are not interested in the dramatic presentations of AI including its replication of human interaction.

How can AI help my business be more profitable? This is the crucial question and, perhaps, the only question that business is interested in. Having implemented AI projects and run successful businesses in the past and, currently, as Chairman of the Board for *All Point POS*, I can tell you with complete certainty that business is extremely keen to know how AI will help solve problems, create profit, support critical thinking, and enable the business to provide customer value. Business is also very keen to remain lean and agile – important business characteristics that are developed with AI.

Way back in 1993, I was a member of a four-person team that implemented the first real-time Neural Network fraud detection system for the Credit Card industry at Household Credit Services. Even this early AI implementation gained traction in large returns on investment which lead to increases to the bottom line for companies. Is the bottom line all that counts? Or does customer value what matters most? Are they so intertwined that you cannot separate them?

Customer value is precisely where this book is unique. While the authors have also provided details on various types of machine learnings and statistical techniques, the focus of their work here is on the application of those technologies and techniques to enable businesses to provide and enhance customer value.

Another important differentiator of this book in your hand is the way it handles the important topic of "optimization". Almost all AI literature focuses on mimicking human brains in order to do what humans do – albeit much faster and with increasing accuracy. This book argues for optimization, which is in the realm of re-engineering of businesses rather than automating them. As a result, the chapters in this book deal with business strategies, business process modeling, quality assurance, and cybersecurity. Business Optimization, as argued in this book, is detailed examination of all business functions to ensure that they use AI in order to generate customer value. This book discusses these important topics around holistic transformation to a digital business using data and AI.

In handling the important softer aspects of AI application to business, the authors do a great job in dedicating an entire chapter on ethics, morality, and biases in decision making. The importance of Natural Intelligence (NI) in making decisions and understanding their consequences cannot be overemphasized. The role of personalities in decision making and, eventually, business agility is something I emphasize during my guest lectures to the MBA class at the University of South Florida. I am delighted to note these soft topics are duly discussed in this book.

Apart from my work in the industry, I am also a proud Rotarian. As this gets written, Rotary is on the brink of introducing a seventh area of focus dealing with sustainability and the environment. The importance of AI and Machine Learning (ML) in tackling the challenges of sustainability is beyond doubt. I look forward to the application of discussions in this book in practice to sustainability and many other such areas that are important for the quality of life on the Earth.

Andy Lyman
(Sarasota, Florida, USA)

Andy Lyman is a leader in the software domain – specializing in enterprise Retail Solutions using the Teamwork Retail software. Andy serves as Chairman of the Board for All Point POS – a leader in Retail Point of Sale Technology. He has delivered guest lectures in the University of South Florida on the intersection of Agility and Leadership. Currently he serves as Vice Chair of the Muma College of Business Advisory Council. Andy is the District Governor Elect in Rotary District 6960 in Florida, USA.

Preface

Artificial Intelligence for Business Optimization: Research and Applications is a business book discussing the research and associated practical application of artificial intelligence (AI) and machine learning (ML) in business optimization (BO). AI comprises a wide range of technologies, databases, algorithms, and devices. This book aims for a holistic approach to AI by focusing on developing business strategies that will not only automate but also optimize business functions by giving due credence to processes and human aspects. The overbearing focus of this book is on using AI and ML from a business viewpoint with the key purpose of enhancing customer value. The research elements in this book are also described from a practitioner's viewpoint. Crucial issues in BO, associated with governance risks, privacy, and security, are addressed in this book to ensure compliance of AI/ML applications from a business viewpoint. Readers should find the discussions in this book direct and practically applicable in their work environment. Researchers will find many ideas to explore further in the applications of AI to business.

The application of AI in business requires a thorough understanding of technology, business, and people issues. Most contemporary AI literature focuses primarily on technologies and associated analytics. This book gives the business primary importance with AI in balance with business decision-making. This book fills the crucial gap existing in the current literature on AI around holistic and strategic application of AI to BO. The nuances of risks and challenges encountered in the transformation of various business functions and corresponding business processes are neatly outlined. The application of AI to not only automate but also optimize business processes based on actionable insights is discussed in this book. This book prepares the reader to apply AI in BO on an ongoing basis.

This book provides substantial discussions for budding researchers who are exploring the industrial applications of AI. This book is also a potential textbook for higher-degree classes in AI and business. The authors have combined their research expertise with practical experiences and contemplations around key topics such as data analytics, machine learning, Big Data, cybersecurity, and sustainability. This book is replete with practical examples that make it easy to understand the concepts and apply them in practice.

This book has direct use for leaders strategizing for BO. The challenges in BO are also highlighted based on the practical experiences of the authors in the industry.

Bhuvan Unhelkar (USA)
Tad Gonsalves (Japan)

Readers

This book will be of immense value to the following readers:

a. Practitioners (consultants, senior executives, decision-makers) dealing with real-life business problems on a daily basis, who are keen to develop systematic strategies for the application of AI/ML/BD technologies to business automation and optimization
b. Practitioners keen to provide and increasingly enhance customer value
c. Researchers who want to explore the industrial applications of AI, machine learning, and Big Data that will reduce the risks of these applications and provide increasingly more value to business
d. People responsible for making policies and establishing governance-risk-compliance (GRC) within and outside an organization, in the industry, and also globally to ensure sufficient security and privacy of data and corresponding AI applications
e. Workshop presenters and participants – typically from the industry – attending a two-day event in a very practical setting (see the outline of a two-day workshop based on this book, below)
f. Instructors and students of a higher-degree course/subject in a university setting

CHAPTER SUMMARIES

Starting with an understanding of AI, ML, and BI, this book develops the idea of utilizing Big Data (BD) analytics for optimized business decision-making. The reader is updated with crucial concepts of the range of ML approaches that handle BD and how to overcome the risks in implementing these approaches. This book contains innovative and entirely new ideas around dynamic learning that have not been discussed anywhere else in the literature on AI. The following is a statement on each of the 11 chapters in this book:

Chapter 1: Artificial intelligence and machine learning: Opportunities for digital business sets the tone for business optimization. and focuses on the business opportunities and customer value.

Chapter 2: Data to decisions: Evolving interrelationships outlines the framework to *Think Data* and how data evolves into decisions.

Chapter 3: Digital leadership: Strategies for Adoption underscores the importance of leadership and strategies in digitizing business using AI.

Chapter 4: Statistical understanding of machine learning types: AI and ML in the business context deals with the statistical algorithms of ML.

Chapter 5: Dynamicity in learning: Smart selection of learning techniques develops the concept of dynamically changing requirements and solutions in ML.

Chapter 6: Intelligent business processes with embedded analytics focuses on the business process aspect of optimization, including modeling and reengineering of processes.

Chapter 7: data-driven culture: Leadership and change management for business optimization underscores the importance of developing an approach to change, which is inevitable in all functions of a business as it optimizes.

Chapter 8: Quality and risks: Assurance and control BO deals with the important topic of quality in the use of Big Data and AI for optimization.

Chapter 9: Cybersecurity in BO: Significance and challenges for digital business draws attention to the importance of security in the use of AI in optimization.

Chapter 10: Natural intelligence and social aspects of AI-based decisions aims to balance the inexplainability of AI with NI for value generation and risk reduction.

Chapter 11: Investing in the future technology of self-driving vehicles: Case study Shows an example of how AI is used in autonomous vehicles.

MAPPING BOOK TO A WORKSHOP

The material in this book is presentable in varying formats. These include:

- A two-day practical training course or a workshop that can be delivered in public or in-house (customized) format to industrial participants (See Table I.1).
- A one-semester, 15-week, university course
- A distance-learning format wherein the assessments, case studies, etc. are based online. Here is a potential mapping of this book to the workshop

Table 1.1 Mapping of the Chapters in This Book to a Two-Day Workshop.

Day	Session	Presentation and Discussion Workshop Topic	Relevant Chapters	Comments
1	8:30–10:00	Introduction to AI, ML, BD, and corresponding business challenges	1	Key concepts and terms are introduced. Significance of business applications is highlighted.
	10:30–12:00	Developing business strategies for optimization	3	Holistic strategy development for AI/ML/BD applications is outlined.
	1:30–3:00	Data to decisions pyramid	2	Evolution of data utilization in decision-making is discussed.
	3:30–5:00	Taxonomy of machine learning and application in business	4	Comprehensive understanding of various ML types, their research-based relevance, and their application in business automation and optimization is discussed.
2	8:30–10:00	Dynamic learning; optimizing business processes	5, 6	Embedding data-driven analytics in business processes and their dynamicity is discussed.
	10:30–12:00	Cultural issues in AI applications; superimposing natural intelligence; quality, security and privacy (GRC)	7–9	"Soft" issues in AI applications to business include quality, security, privacy, natural intelligence, and so on. These are discussed here.
	1:30–3:00	Understanding and working through a case study	3, 4, 5, 11	Participants move to a workshop format and develop a business optimization strategy with AI.
	3:30–5:00	Handling practical challenges and risks associated with BO	All	Participants share and outline their business strategy in optimization; and discuss and present their thoughts, issues, and challenges

Figures

All figures in this book are based on consulting work performed by the authors. They are owned and are produced here by the original authors.

Acknowledgments

Warren Adkins

Anurag Agarwal

Aurilla Aurelie Arntzen

Walied Askarzai

Josh Baker

Milind Barve

Yi-Chen Lan

Colleen Berish

Madhulika Bhatia

Bhargav Bhatt

Steve Blais

Asim Chauhan

Vivek Eshwarappa

Diego Felipe

Abbass Ghanbary

Ram Govindu

Naman Jain

Haydar Jawad

Jean Kabongo

Sunita Lodwig

Andy Lyman

Masa K. Maeda

Mohammed Maharmeh

Javed Matin

San Murugesan

Girish Nair

Suresh Paryani

S.D. Pradhan

Mukesh Prasad

Trivikrama (T.V.) Rao

Abhay Saxena

M.N.Sharif

Sanjeev Sharma

Keith Sherringham

Devpriya Soni

Prince Sounderarajan

Amit Tiwary

Bharti Trivedi

Authors

Prof. Dr. Bhuvan Unhelkar (BE, MDBA, MSc, PhD, FACS) has extensive strategic and hands-on professional experience in the Information and Communication Technologies (ICT) industry. He is a full professor and lead faculty of IT at the University of South Florida (USF) and is the founder and consultant at *MethodScience* and *PlatiFi*. He is also an adjunct professor at Western Sydney University, Australia, and an honorary professor at Amity University, India. His current industrial research interests include AI and ML in business optimization, Big Data and business value, and business analysis in the context of Agile. He holds a Certificate-IV in TAA and TAE, Professional Scrum Master – I, SAFe (Scaled Agile Framework for Enterprise) Leader, and is a Certified Business Analysis Professional® (CBAP of the IIBA).

His areas of expertise include:

- Big Data strategies: *BDFAB* – with an emphasis on applying Big Data technologies and analytics to generate business value
- Artificial intelligence and business optimization
- Agile processes: *CAMS* – practical application of composite Agile to real-life business challenges not limited to software projects
- Business analysis and requirements modeling – use cases, BPMN, and BABOK; helping organizations up-skill and apply skills in practice
- Software engineering – UML, object modeling; includes undertaking large-scale software modeling exercises for solutions development
- Corporate Agile development – up-skilling teams and applying Agile techniques to real-life projects and practice
- Quality assurance & testing – with focus on prevention rather than traditional detection
- Collaborative web services – SOA, Cloud; upgrading enterprise architectures based on services, including developing analytics-as-a-service
- Mobile business and green IT – with the goal of creating and maintaining sustainable business operations

His industry experience includes banking, finance, insurance, government, and telecommunications where he develops and applies industry-specific process maps, business transformation approaches, capability enhancement, and quality strategies.

He has authored numerous executive reports, journal articles, and 22 books with internationally reputed publishers, including *Outcome Driven Business Architecture* (Taylor and Francis/CRC Press, USA, 2019), *Software Engineering with UML* (Taylor and Francis/CRC Press, USA, 2018), and *Big Data Strategies for Agile Business* (Taylor and Francis/CRC Press, USA, 2017). Cutter *Executive Reports* (Boston, USA), including *Psychology of Agile* (two parts), *Agile Business Analysis* (two parts), *Collaborative Business & Enterprise Agility*, *Avoiding Method Friction*, and *Agile in Practice: A Composite Approach*. He is also passionate about coaching senior executives; training, re-skilling and mentoring IT and business professionals; forming centers of excellence; and creating assessment frameworks (SFIA-based) to support corporate change initiatives.

Dr. Unhelkar is an engaging presenter delivering keynotes and, training seminars and conducting workshops that combine real-life examples based on his experience with audience participation and Q&A sessions. These industrial training courses, seminars, and workshops add a significant value to the participants and their sponsoring organizations because the training is based on practical experience, a hands-on approach, and accompanied by ROI metrics. Consistently ranked highly by participants, the seminars and workshops have been delivered globally to business executives and IT professionals in Australia, USA, Canada, UK, China, India, Sri Lanka, New Zealand, and Singapore. He is the winner of the IT Writer Award (2010), Consensus IT Professional Award (2006), and Computerworld Object Developer Award (1995). He also chaired the *Business Analysis Specialism Group* of the Australian Computer Society.

Dr. Unhelkar earned his PhD in the area of "object orientation" from the University of Technology, Sydney. His teaching career spans both undergraduate and master's level wherein he has designed and delivered courses, including *Global Information Systems*, *Agile Method Engineering*, *Object-Oriented Analysis and Design*, *Business Process Reengineering*, and *New Technology Alignment* in Australia, USA, China, and India. Online courses designed and delivered include Australian Computer Society's distance education program; the M.S. University of Baroda (India) Master's program; and, currently, *Program Design with the UML* and *Mobile App Development* at the University of South Florida Sarasota-Manatee, USA.

At the Western Sydney University, he supervised seven successful PhD candidates, and published research papers and case studies.

Professional affiliations include:

- Fellow of the Australian Computer Society (elected to this prestigious membership grade in 2002 for distinguished contribution to the field of information and communications technology), Australia
- IEEE Senior Member, Tampa Chapter, USA
- Life member of the Computer Society of India (CSI), India
- Life member of Baroda Management Association (BMA), India
- Member of Society for Design and Process Science (SDPS), USA
- Rotarian (Past President) at Sarasota Sunrise club, USA; Past President Rotary club in St. Ives, Sydney (Paul Harris Fellow; AG), Australia
- Discovery volunteer at NSW parks and wildlife, Australia
- Previous TiE Mentor, Australia

Previous Books by Dr. B. Unhelkar, published by CRC Press (Taylor & Francis):

Hazra, T., and **Unhelkar, B.**, (2020), *Enterprise Architecture & Digital Business,* CRC Press, UK.

Sharma, S., Bhushan, B., **Unhelkar**, B., 2020, *Security and Trust Issues in Internet of Things: Blockchain to the Rescue*, Edited, CRC Taylor and Francis Group, USA.

Tiwary, A., and **Unhelkar, B.**, (2018), *Outcome Driven Business Architecture,* CRC Press, (Taylor and Francis Group/an Auerbach Book), Boca Raton, FL, USA. Co-Authored.

Unhelkar, B., (2018), *Software Engineering with UML,* CRC Press, (Taylor and Francis Group /an Auerbach Book), Boca Raton, FL, USA. Authored, Foreword Scott Ambler. ISBN 978-1-138–29743-2.

Unhelkar, B., (2018), *Big Data Strategies for Agile Business,* CRC Press, (Taylor and Francis Group/an Auerbach Book), Boca Raton, FL, USA. Authored ISBN: 978-1-498–72438-8 (Hardback), Foreword Prof. James Curran, USFSM, Florida, USA.

Unhelkar, B., (2013), *The Art of Agile Practice: A Composite Approach for Projects and Organizations,* CRC Press, (Taylor and Francis Group/an Auerbach Book), Boca Raton, FL, USA. Authored ISBN 9781439851180, Foreword Steve Blais, USA.

Unhelkar, B, (1999), *After the Y2K Fireworks: Business and Technology Strategies CRC Press*, Boca Raton, USA; July 1999; Total pages: 421. (Foreword by Richard T. Due, Alberta, Canada).

Prof. Dr. Tad Gonsalves (MSc, PhD) is a full professor in the Department of Information and Communication Sciences, Faculty of Science and Technology, Sophia University, Tokyo, Japan. His research areas include bio-inspired optimization techniques and the application of deep learning techniques to diverse problems like autonomous driving, drones, digital art, and computational linguistics. Dr. Gonsalves has published nearly a hundred papers in international conferences and journals on areas such as knowledge management and engineering, meta-heuristic and bio-inspired optimization, fuzzy systems, machine learning, autonomous driving, and other AI-related applications. He holds a BSc in Theoretical Physics, MSc in Astrophysics, and earned his PhD in Information Systems from Sophia University, Tokyo, Japan. Dr. Gonsalves has published nearly a hundred papers in international conferences and journals on areas such as knowledge management and engineering, meta-heuristic and bio-inspired optimization, fuzzy systems, machine learning, autonomous driving, and other AI-related applications.

His research laboratory in Tokyo (https://www.gonken.tokyo/) specializes in applications of deep learning and multi-GPU computing.

Previous Book by Dr. Tad Gonsalves (2017), *Artificial Introduction: A Non-Technical Introduction*, Sophia University Press, Tokyo, Japan.

Chapter I

Artificial intelligence and machine learning

Opportunities for digital business

ARTIFICIAL INTELLIGENCE IN THE CONTEXT OF BUSINESS

Artificial intelligence (AI) is precisely that: it is non-real intelligence that is demonstrated by machines (computers) as they learn and mimic human (natural) intelligence. AI helps businesses achieve their goals based on the parameters provided to its algorithms that analyze vast and relevant data. Success with AI requires an understanding of the business. This is one of the crucial differentiators in a strategic approach to business optimization (BO). Analytics in Big Data are important but not without a proper understanding of business.[1]

AI is understood here as a combination of systems, processes, algorithms, and techniques to analyze large, complex, and fast-moving data. The purpose of such analytics is to identify trends and patterns that will help extract insights from the data. Decisions based on the insights are as close to human decision-making as possible and are continuously and iteratively improving on the results. Over decades, thinkers like Davenport[2] have called AI as "the most important general-purpose technology of our era with wide ranging applications." This discussion narrows the potentially wide-ranging applications of AI to the one that enables businesses to provide "value" to their customer. An optimized business provides this value in the most efficient and effective way using the tools and techniques of AI. As a result, the business becomes agile. Agility is an important characteristic of business in the digital era.[3] BO is thus a strategic application of AI and ML using Big Data in order to provide value to customers with agility. BO is not limited to data and processing. BO uses AI to expand to reach the outer edges of the business wherein it "spots" the customers, understands their needs, personalizes the offerings, and continuously enhances the products or services. AI also helps the business handle governance, privacy, security, and compliance requirements. The keywords of importance are *systems, processes, algorithms, techniques, analyze, extracting, insights, large, complex, fast-moving data, support, decision-making, value, efficiency, effectiveness, agility,* and *customer.* Each of these words has a meaning that is specific to the discussion on AI and BO. This book is a journey into these concepts of AI and their application to BO.

Artificial intelligence (AI) and machine learning (ML) as enablers of business optimization (BO)

AI brings together a wide range of technologies, databases, algorithms, and devices as potential enablers of BO. AI provides the necessary optimization capabilities. AI on its own, however, is not BO. BO acknowledges the need to strategically apply AI to business functions. Furthermore, AI is presumed to bring about automation. Contemporary digital business strategies are mainly geared towards automation,[4] which is not the same as optimization. Optimized businesses are, by necessity, digital businesses but the reverse may not be true. This is so because digital businesses may be automated with AI and ML but not necessarily optimized. Optimization goes further than automation by not merely mimicking the existing processes but ensuring business goals with efficiency and effectiveness. Optimization in this discussion is specifically delineated from automation as a separate and dedicated business initiative.

Big Data provides AI and ML with the necessary range, depth, and variety of data that can be analyzed in order to produce excellence in customer value. BO is not an isolated activity or a project in an organization. BO starts strategically, at the board level, with the examination of the entire business, the environment in which it operates, its functional and structural parameters, and its challenges and opportunities. AI is the enabler to revamp the business based on a long-term view of the business and the environment in which it operates. BO covers people, processes, technology, budgets, security, and quality issues for business and its collaborating partners.

BO initiative starts by studying each business objective and function keeping AI capabilities and constraints in mind. As a result, BO minimizes the risks and maximizes the agility of a business. The decision variables, the objective functions, and the corresponding constraints within a business process are carefully quantified in BO keeping technology restraints in mind. The decision variables serve as inputs to the AI optimization algorithms (e.g., evolutionary computation and swarm intelligence), which can optimize even large-scale business problems in a reasonable amount of time.

Subjective elements in BO

While BO capitalizes on the AI algorithms, it also gives due credence to the nonquantifiable aspects of AI. These unquantified concepts are "subjective." The acknowledgment and insertion of the subjective element in the optimization of business processes are based on human or natural intelligence (NI).

NI is an important part of holistic BO, and it is discussed in Chapter 10. Other issues such as lack of sufficient explainability of AI[5] are discussed in later chapters.

Machines are very good at mimicking the human functions. Gaming, autonomous driving, robotic surgery, and medical diagnostics are all capabilities that machines can handle well. Computers are phenomenally better at undertaking routine tasks, and the more routine they are, the better is the automation. Increasingly complex tasks require correspondingly complex algorithms which are coded by humans.

Machines, however, lack the cognitive capabilities inherent to humans. As Finlay[6] puts it, "True AI is about much more than just pattern recognition and prediction." He further questions "Is there some additional (as yet unknown) element required for human-like intelligence and self-awareness which can't be replicated via computation alone?"[7] There is no inkling of that question being answered in the near future. Ada Lovelace expresses this succinctly:

> The Analytical Engine has no pretensions whatever to originate anything. It can do whatever we know how to order it to perform. It can follow analysis; but it has no power of anticipating any analytical relations or truths. Its province is to assist us to making available what we are already acquainted with.[8]

The "learning" and "problem-solving" capabilities of AI are programmed by developers. Decision-making uses the analytical insights but is still carried out by people. The quality and goodness in decisions are judged by their consequences which may be subjective and beyond the scope of AI. NI ("humanization") is a positive influence on AI-based decision-making.

Agility in BO

Agility is an important business characteristic. An agile business is a flexible business that can change according to the changing needs of the users and the environment. Automation, optimization, and humanization are considered together in holistic BO to provide agility to business and enhance customer value. The variances and nuances of business functions, their automation, and their optimization occur *on an ongoing basis*. AI, ML, DS, and Big Data are considered as the technologies that enable optimization.

Data provides the backbone of the systems and processes of a digital business. Analytics support the optimization by extracting value. Data, however, is exploding. The blinding speed, mounting volumes, and vast geographical reach of digital data result in Big Data. Systems, processes, and techniques that make sense of this Big Data have to be continuously evolving. Business and technologies need agility in order to keep pace with Big Data. Agility in design and implementation of analytics enable not only processing of data but also "remembering" the algorithm execution in an iterative and incremental manner. This ability to "learn" from an execution of the algorithm for varied sets of data is ML. ML design, implementation, and testing require agility. Agility during BO occurs in both the business and the solution space.

AI also personalizes and customizes the visualization of analytical insights based on specific needs of the users, their collaborative choice, and their level of interest. The flexibility in viewing the analytics across business boundaries and over many devices is a part of business processes. AI also maintains consistency and currency of visualization across collaborative businesses.

Collaboration in BO

Another important characteristic of an agile digital business is its extension and reach to other supporting businesses. Multiple businesses are able to collaborate with each other in a distributed and federated environment through digital technologies. Big Data technologies enable the generation and assimilation of dynamically changing contents from a variety of internal or external data sources. AI and ML technologies merge contents developed and published by different sources as appropriate to the required analytics. Decision-makers in collaborative enterprises use data and analytics from a variety of sources on the cloud that may not be owned by them.

Granularity in BO

A crucial factor in providing efficiency and effectiveness in AI is the concept of granularity in data analytics.[9] The AI technologies enable the analytics to drill down to the finest level of detail.[10] This analytical capability is important because of the large, complex, and fast-moving (high-velocity) data together with the demand for the analytical results in a short time span. ML is used to achieve increasingly finer levels of granularity – such as identifying the precise product for a customer at a particular point in time, narrowing down potential areas of fraud and money laundering, and enabling emergency services (e.g., ambulance and fire) to position themselves for rapid responses on certain days or at certain events. Granularity is thus important in AI. Strategic use of AI requires the incorporation of data-driven decision-making that is agile, collaborative, and right granular. Data-driven decision-making requires a thorough understanding of customers, their personalized needs, capabilities of the business, and staff training. The need and the opportunity to reengineer business processes across all functions of the business could not have been higher.

The technical-business continuum

Figure 1.1 positions data science (DS) on a technical-business continuum. Deep learning (DL) is on the technical side of the continuum, followed by

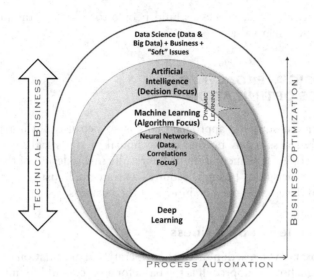

Figure 1.1 DS, AI, ML, neural networks, and DL (ML is a subset of AI).

neural networks, ML, AI, and the overall DS domain. These technologies increasingly move towards the business side as shown in Figure 1.1. The core DS technologies provide the basis for the automation of business processes. Optimization, however, is treated as more than automation and is the key business goal in this discussion.

The increasing layers from DL to business and "soft" issues, shown in Figure 1.1, play an important role in the optimization effort. Dynamic learning (DL), shown in Figure 1.1 across the layers of ML and AI, is an innovative approach to provide elasticity in the selection and application of the ML algorithms in business. This dynamic learning is an effort to automate the optimization process and is discussed in Chapter 5.

AI technologies are disruptive because they have the potential to dramatically change the macro- and micro-business environments. While ML can help harvest knowledge, as the data gets bigger and arrives faster, predictive analytics solutions based on DL come into play.[11] DL techniques are essential to support predictive analytics and aid in knowledge discovery especially in the vast repositories of Big Data repositories. DL uses supervised and unsupervised approaches to learn multilevel representations and features in hierarchical architectures for the tasks of classification and pattern recognition.[12]

AI also provides new and interesting opportunities for resilience and recovery functions of an organization. AI, through its ability to analyze vast amounts of data, can describe, predict, and even prescribe disruptions to business functions. This is so because AI, under the umbrella of DS, utilizes data (the most valuable asset of an organization) to produce insights.

BO puts technologies, processes, and people together for the purpose of providing customer value.

STRATEGIC APPROACH TO BUSINESS OPTIMIZATION

BO demands a strategic approach. This is so because BO impacts every aspect of business – way beyond the technical or statistical aspects of AI. In fact, a proper implementation of BO results in a redesign of the business itself. This redesign of business requires a strategic, holistic approach that is based on developing the capabilities of the business.

BO as a redesign of business

Contemporary AI is used in tactical or operational automation of business. The strategic holistic approach is the basis for a successful BO initiative. BO with AI/ML impacts all functions of a business. For example, accounting, marketing, sales, inventory, and HR are all impacted by the application of AI in business. BO is effectively reengineering the entire business.

BO aims for an organic, lean, and agile business. BO is also an ongoing activity with iterative and incremental changes. Therefore, every aspect of the business changes due to BO. The processes change due to the use of analytics in arriving at faster and more accurate decisions. The people also undergo change – both the customers and the users. The customer discovers new and productive ways of interacting with the business, and the staff learn to provide greater value and more personalization in their offerings. Astute leadership with a good understanding of change management is crucial in the success of this redesign of business due to AI.

Developing a BO strategy

A good BO strategy keeps the business organization at the center. In doing so, the customer also comes to the fore because that is where the maximum impact of BO is felt – in maximizing the customer value. A BO strategy is developed based on due consideration to the following questions:

- What is the type of the organization? Is it a product, service, or entirely online organization? Organizational perceptions and expectations of customers differ for each type of organization.
- What are the internal challenges faced by the business? These challenges are prioritized in developing a BO strategy.
- How to differentiate the short-term and long-term challenges faced by the business? The short-term, tactical-operational challenges are more suitable for automation, whereas long-term challenges are optimization.

- What are the business risks in the application of AI in data-driven decision-making? How will the customer perceive an optimized business?
- What are the current risk management approach and the likely approach to handling risks?
- How can AI be applied to businesses while it is in operation? The current business will continue during the adoption process.
- What is the type and size of data used in the business currently? And how much of this data is currently used in business decision-making?
- How matured are the current processes of the business? The more matured the processes, the easier it is to embed and test AI.
- How can practical experiences in AI, ML, and Big Data be shared across the business and the industry in a manner that highlights the challenges and risks associated with these technologies in business?
- What do the senior business leaders know and understand in terms of AI, ML, and Big Data? Without a proper and technical understanding of AI and ML, the leadership functions and strategies for BO will suffer.
- How do senior leaders view AI as capabilities that need people and people skills? How to source (recruit) DS talents?
- When, where, and how can users/staff be trained in data-driven decision-making?
- What are the potential risks of not adopting AI? Which crucial opportunities will be lost if AI is not adopted?
- What are the issues in using known statistical techniques in decision-making? Technical? Ethical? Business? AI is usually not explainable although the DS community is striving to make it so.
- How can BO aid and support governance-risk-compliance (GRC)?
- What is the depth of cybersecurity in the organization and how will it change with AI? How can the networks, sensors, and data analytics be applied for cybersecurity intelligence?
- Does the business know its customers sufficiently enough to provide for their personalized and timely needs? This requires a uniform master data management initiative already in the organization. Analytics on multiple copies of the same data can lead to chaos in decision-making.
- Does the business understand what the customers are really looking for? Costs, personalization, privacy, and security are some of the factors customers are looking for.
- What will cause customers to switch to a competitor? Which type of data and analytics can throw light on customer behavior?
- How can data help the business understand its competition better? And how fast is the data able to make the prediction of changes?
- Where are the potential new customers? What is their demographic? How will the demographics of customers change? Locally, globally? What are the sources of data for these customers (e.g., alternative data, social data)?

- What are the product-customer matches? Least successful?
- Does the business understand its staff? What is the satisfaction level of the staff? Who are likely to leave the employment and move elsewhere?
- What are the existing and future quality initiatives? How will those initiatives impact and be impacted by BO?
- What are the costs associated with securing the sources, storages, and retiring of data?

Capabilities in BO

The redesign of a business with AI is actually a review and redesign of business capabilities. Capabilities are meant to provide the businesses with configurable ability to achieve their strategic goals.[13] ML-based capabilities are able to help the organization to optimize its processes and equip its people to achieve its strategic goals. These capabilities, equipped with AI, are continuously aligned with the desired business outcomes.[14] Alignment of AI-enabled capabilities with the business goals ensures that the organization is not wasting time, effort, and money in developing solutions that are skewed from the business goals. Business architectures are invaluable in enabling the alignment of capabilities with goals. Such aligned capabilities also ensure that the business is well prepared for the inevitable disruptions.

Thus, an important responsibility of AI usage in business is to equip the business with capabilities to handle disruptions. Such capabilities also enhance the leadership's capabilities to handle the challenges arising from disruptions and changes.

AI, BIG DATA, AND STATISTICS

AI, Big Data, and statistics overlap each other within DS. Figure 1.2 summarizes various terms in DS and their meanings. As shown in Figure 1.2, AI interacts with BD through the ML algorithms. Statistics is devoted to the analysis of data using conventional algorithms. The interaction of AI with statistics gives rise to business intelligence (BI), which is one of the recent disciplines found in business. The interaction of AI and statistics with BD produces analytics, which is essential for decision-making.

Data, science, and analytics

Data science is a broad-ranging term that represents the technologies and analytics of data.[15] Data includes Big Data and the technical, business, and social considerations vital for extracting value. Therefore, DS deals with data networks, machines, end-user devices (IoT), customers, users, security and privacy, algorithms, and "soft" topics such as ethics and law. DS starts

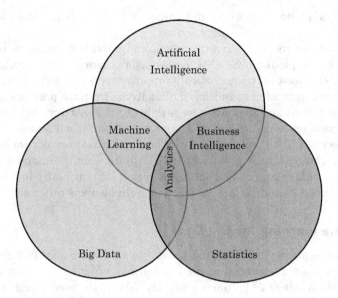

Figure 1.2 Data science – a summary of various terms in DS and their interrelationship.

with the basic premise that data has business "value" hidden in it. This premise leads to knowledge discovery and data mining, which have been investigated by various authors.[16,17,18] Data mining, analytics (statistics), process modeling, ML, parallel processing, and data management are applied to discover the value. The application of analytics to these data is the main step of arriving at insights. DS is important to the business leadership and the strategic direction it sets for the organization. DS also aspires to convert data into actionable knowledge.

Evolution of data to actionable knowledge requires a specialist discipline that includes the study of data, its characteristics, its context in analytics, and its value in business agility. DS is interdisciplinary because it includes disciplines within the organization and collaboration with many cross-functional teams.

What and why of ML?

ML algorithms are able to perform numerous crucial digital functions in a business that are beyond human capabilities. ML uses statistical algorithms to spot trends in large-volume and high-velocity data. ML is further understood as machines (computers) programmed to learn iteratively and incrementally from the data provided and analytics conducted. ML algorithms modify their analysis based on the dynamically changing input. ML ingests analytical results from one iteration and improves the prediction in the next iteration. ML also creates new and unique insights by dynamically learning from the data in the earlier iteration.

ML provides tools to learn from data and provide data-driven insights, decisions, and predictions.[19] ML is seen as an algorithm that builds computer applications that automatically improve with past experience.[20] ML (a) improves the accuracy and speed of analytics from the previous iteration and (b) develops and suggests new insights to the user incrementally. ML applications improve the speed, accuracy, and the reach (extent) of business processes. This enhancement to business processes occurs with the help of techniques such as (a) scorecards, (b) decision trees, and (c) neural networks. ML enhances business agility as it incrementally improves the analytical results and facilitates the changes in business processes.

Machine learning for Big Data

Big Data is so big that making sense out of it is humanly impossible. Therefore, there is the need to use machines in order to make sense of the data. This is where ML comes into play. ML is the mechanism to make sense out of Big Data. Machines can process data and store the results. In subsequent iterations of processing data (including a suite of data), machines utilize results from the previous iteration. Thus, data and its corresponding results are stored and reused in an iterative and incremental manner – leading to various types of "learnings" for machines.

Simulating human decision-making and storing the results from those simulations provide increasing accuracy in processing high-volume and high-velocity data. At the root of those decisions, however, is the algorithm and data that is put together by human intelligence.[21] Therefore, utilizing Big Data in decision-making requires ML.

Automation with ML

The size and complexity of Big Data are such that humans cannot correlate all datasets on their own. ML is focused on creating correlations between data points. These correlations can be established without explicit instructions. Data points relate with each other *automatically,* opening up opportunities to create novel analytical insights. ML can learn iteratively (an agile characteristic) within a set of permissible rules. While Big Data analytics create insights, those insights are still limited to the effort and imagination of the individuals undertaking those analytics. Big Data presents a limit to the effort and imagination. ML models are built through iterative and interactive learning that can almost be described as meta-programming. This is automation in ML developments. Automated ML-based model is based on the following questions:

- How to enable data points to relate to each other so that they form a sensible mosaic that will be of interest in the analytics?

- What should be the guiding (or limiting) parameters surrounding a data point to enable it to seek and connect with another data point?
- How can a connection between two data points provide a feedback or "learning" mechanism for the background algorithm?
- How many properties of a data point will be optimal to enable it to create new (and sensible) links?
- What is the business value of such automated ML-based interconnections? How will the interconnections interest the decision-makers?

Applying ML in practice for BO

Cheaper data storage, distributed processing, more powerful computers, and the analytical opportunities available have dramatically increased the opportunities to apply ML in business systems.[22] ML techniques are applied to a wide range of complex problems.[23] ML is embedded in tools that express the domain of expertise.[24] Practical application of ML requires a good set of assumptions. These business and technical assumptions enable easier and faster statistical learning processes for the machines. ML algorithms based on statistical analysis provide mechanisms for software applications to predict outcomes without being explicitly programmed.[25] ML is ideally applicable for tasks that are far too complex to program and where there is a need for the systems to learn and improve based on previous learning patterns through some "experience." ML is invaluable in a changing environment.[26]

Business intelligence

BI makes use of AI and statistical techniques in business decisions. BI is thus an application of data-analytic thinking[27] and data management skills. Applying the results from the analytics in practice requires business knowledge. BI, thus, depends on DS domain knowledge of the industry. Banking,[28] industrial quality control,[29] predictive maintenance, finance, insurance, telecom, medical, education, and even elections[30] are such example domains. In each case, business knowledge is important in developing data-driven strategies for BI. Following are some examples where AI and statistical analytics are combined with business domain knowledge to produce BI:

- Undertaking detailed analytics to suggest, with a degree of confidence, pricing of an airline ticket. This prediction requires knowledge of the airline pricing strategies, history of pricing, trends in the industry, suitable data, and corresponding analytics. The optimization of this predictive process requires knowledge of statistical modeling, data sources on airline prices, and an understanding of the sectors, schedules, and yields.

- Analyzing a bank's internal enterprise systems data and combining the results with demographic meta-data in order to identify potential loan defaulters based on knowledge of credit risks[31] and associated regulations. Knowledge of the banking and finance domain and the inner nuances of loan defaults is required. Simply sourcing a FICO score[32] (a three-digit number predicting the likelihood of loan repayment) may not be enough, and the superimposition of NI together with how the business operates is considered essential for customer value.
- Predict the risk of credit card fraud based on a wide range of micro (unidentifiable individual) and macro (group demographic) parameters and the personal credit market. This activity requires knowledge of financial fraud detection in addition to the knowledge of credit data and its analysis. Additionally, ethics and moral values also play a role in arriving at final decisions on assessing credit frauds using AI.
- Predict the capacity for storing agricultural produce by bringing together weather, soil, and economic data – requiring a combination of knowledge in multiple disciples of weather, agriculture, economy, and supply chain logistics (for transportation). These kinds of predictions require experts from multiple disciplines with the ability to create multiple "what-if" scenarios with the help of AI.
- Explore the revenue trends and relate them to customer turnover for a hotel chain. Developing this data pattern requires knowledge of the hospitality domain and how predictions on occupancy and turnover are made. The business systems used by the hotel chains include AI and NI.
- Preparing production schedules and support logistics for the delivery of goods from a manufacturing organization requires knowledge of production scheduling together with the predictive aspect of data analysis and the associated supply chain.
- Developing a promotion strategy in a democratic election process based on fine granular analytics on keywords and their relationship to voter demographics. Such promotion needs an understanding of the political and voting process as much as analytical process.
- Facilitating organization of communities around common data and analytical interests (e.g., buying groups, political groups, and environmental groups). This requires an understanding of how communities are formed, what sustains them, the risks associated with their formation, and the value they provide.

ML TYPES IN BO

Learning occurs in the enterprise systems in different ways. The learning paradigms for machines are most commonly grouped into three categories of ML[33]: supervised learning, unsupervised learning, and reinforcement learning

(RL).[34] Additionally, these techniques are combined in DL. These learning paradigms are introduced next. Further detailed discussion with examples is in Chapters 4 and 5.

Supervised learning

Supervised learning predicts future target variables, whenever the values of the input attributes are known and have a sufficient amount of accurate data. Supervised learning algorithms predict a value based on existing historical data using regression and classification. Supervised learning maps feature attributes to the target class and then compares the prediction of the target class to the ground facts.

When the target class is a set of discrete values, it is a classification task; when they are continuous numerical values, it is a regression task.[35,36] Examples of classification tasks are whether a customer will remain loyal to the company or not.[37] The supervised learning algorithm that takes sample input and corresponding expected output and "learns" from that relationship. As a highly simplified, conceptual example, consider the sum of two numbers, "2 and 3" with the result "5". The algorithm "learns" to add any two given numbers; so, when provided with another set of numbers, say 4 and 6, the result is computed as 10. Alternatively, the algorithm can be "taught" to find the missing number if only one input and the result is provided (e.g., a number "2" is provided and a result "5," then the missing number is "3").

Supervised learning plays a significant role in BO as systems from multiple, collaborating parties can be *taught* based on previous decisions to undertake similar decisions in a shorter time. Systems can also be taught to flag exceptions in decisions for human intervention. The data representation in supervised learning needs higher quality in order to ensure good performance on the learnt patterns.

Unsupervised learning

Unsupervised ML algorithms find patterns in data without having any prior knowledge of the dataset.[38,39] As with supervised learning, the quality of data representation is also important here. Unsupervised ML algorithms learn from the feature space and classify data points into clusters such that the points in a given cluster are similar to one another and different from the points in the other clusters. Iterations and increments finesse the learnt patterns and provide increasingly improving performance from unsupervised learning algorithms.

Continuing with the overly simplified example of the previous section, the algorithm for analytics is not specified in detail but the three numbers are simply made available – "2, 3, and 5." The algorithm develops its own logic in order to discern that when 2 and 3 are added, the result is 5. This "learning" can be verified over a large set of training dataset running into millions (or billions) of records. Without the technologies that support Big Data, this learning is not possible, as it requires substantial computing power in a distributed architecture. The learning algorithm discovers hidden patterns and constructs its own logic that can help users consider the possibility of new questions. The algorithm is thus "learning to learn" based on the discoveries in initial iterations.

Systems can be exposed to databases containing vast amounts of data and made to come up with *themes* around that data. While the systems themselves may not be able to initially identify what these themes imply, later, iteratively and incrementally, an increasingly well-defined interpretation of the theme emerges. This ability of ML to dive into a vast amount of data that would not make sense to a regular ERP solution is important in order to incorporate intelligence in optimizing business processes.

Reinforced learning

RL is effectively like teaching a dog to learn new tricks. When the dog performs a trick as directed, it is rewarded; when it makes mistakes, it is corrected or penalized. The dog soon learns the trick by mastering the policy of maximizing its rewards at the end of the training session. Robots learning to grip objects, maneuver through hazardous zones, and gently lift patients from hospital beds; self-driving cars learning to drive by observing traffic rules and avoiding accidents; and software programs learning to beat world champions are offshoots of advanced RL algorithms.

The agent (software program engaged in learning) is not given a set of instructions to deal with every kind of situation it may encounter as it interacts with its environment. Instead, it is made to learn the art of responding correctly to the changing environment at every instant of time. When the agent acts on the environment, the action is evaluated by the environment and the agent is either rewarded or penalized. At the same time, the action of the agent also changes the state of the environment and the agent acts again. The series of state-action-reward cycles in a training session makes the agent learn a policy that forgoes instant gratification to accumulate the long-term rewards.

Deep learning

DL is an extension of AI that provides value during BO. Neural networks, which are based on the nervous system of the human brain, are modeled to implement DL. The human brain learns and knows where to apply that learning through multiple layers. DL algorithms mimic these human capabilities in order to solve multilayered business problems.

DL is playing an increasing role in Big Data predictive analytics in particular.[40] DL uses supervised and unsupervised strategies to learn multilevel representations and features in hierarchical architectures for the tasks of classification and pattern recognition.[41]

DL has a broad range of applications such as natural language processing and speech recognition. For instance, DL improves the customer experience by integrating chatbots or conversational AI assistant with enterprise solutions. DL correlates knowledge and information across industries creating new opportunities through intelligent, automated process, services, or products. Chatbots are integrated in various applications like travel, entertainment, health, and education.

Feature engineering

Feature engineering builds features and data representations from raw data for business use. Feature engineering is domain specific as it aims to develop capabilities for the specific business. Automated feature engineering enables the extraction of features relevant to the business with a minimal human intervention. Since automatic feature extraction by deep learning, these algorithms are a lot more than mere automation. DL algorithms in feature engineering have opportunities to optimize the solutions more than automating them. In all other types of ML, features must be provided by domain experts as part of feature engineering.[42]

DIGITAL BUSINESS AUTOMATION AND OPTIMIZATION

Automation and optimization of digital business are goals in using AI, ML, and BD. Digital business is not simply an extension of a physical business. Data is the basis for a comprehensive strategy for BO. The ubiquitous website and a basic mobile app are only a start for digital business. Data provides many strategic opportunities for optimization. Accounting, airline, health, financial trading, education, and sports are examples of domains that have benefitted by strategic use of data. Data-driven business is a continuous evolution of a business into an automating, optimizing, and collaborating phenomena for the digital world.[43]

Value extraction from data

Figure 1.3 shows the relationship between automation, optimization, and the influence of subjectivity (humanization) in extracting value from data. Optimization is not considered independent of automation and

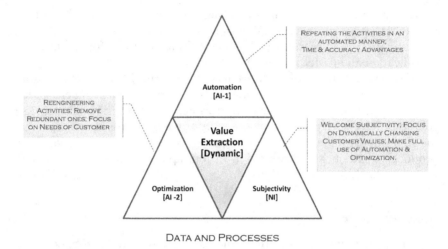

Figure 1.3 Automation–optimization–humanization in value extraction.

humanization. DS enables an efficient execution of routine processes resulting in automation. Automating the routine organizational functions provides cost and time benefits. However, automation with ability to improve on itself is intelligent automation. AI provides substantial inputs for this intelligence in automation.[44] DS needs to handle the subjectivity in value-based optimization. DS provides the following advantages when considered together with automation, optimization, and humanization:

- Insights on inventory, production, and scheduling processes enable the optimization of internal business operations. Some aspects of inventory and production that are routine and well defined may initially be automated. Full advantage in these processes is derived only with optimization.
- Reach wider cross section of customers through digital communications, devices, and communities. While initially this reaching out can be automated, the process has to be fully modeled and redundancies eliminated during optimization.
- Offering personalized services to customers by a detailed understanding of their needs based on the context is not an automated task; instead, it needs a significant humanization input.
- Combine multiple services to offer a wider range of products to customers from a single point based on automated processes that are modeled together for optimization.
- Improve regulatory compliance and auditability of records by proper and formal maintenance of data and using advanced searchability features.
- Enhance the quality of service by not only automating, but also applying optimization to the services.

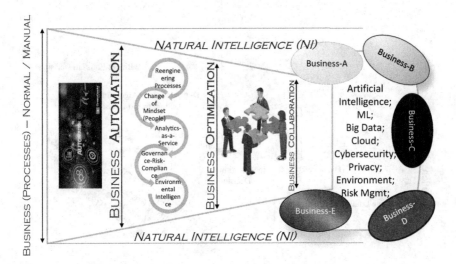

Figure 1.4 Intelligence BO is more than automation, includes NI, and opens up significant collaborative opportunities.

- Enhance the security of processes using AI and enhance the security of AI itself.

Intelligent optimization

Optimization implies the most effective use of resources in a given situation. Automating the existing processes and thereby gaining time and effort advantage is necessary, but not sufficient. Optimization is the critical examination of a business process keeping its ultimate outcome in mind. Therefore, intelligent optimization is more than automation. Figure 1.4 shows the application of AI and NI to business processes. Optimization uses technologies to eliminate redundant activities within a process altogether. While automation is necessarily based on technologies, optimization starts by examining business processes independent of technologies. This examination is followed by analytics, communications, and ongoing critical examination of the goals of a process. Ongoing improvements in the processes and functions of the business in order to serve a business goal is intelligent optimization. The application of NI enables the business to maximize the value it offers to its customers. Collaborations between various businesses enable a widening of offerings to customers and a reduction in risks.

Increasingly complex business situations

As the business evolves through automation and optimization, it utilizes data from routine to complex business processes. Figure 1.5 shows how

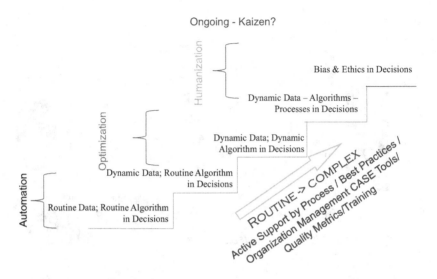

Figure 1.5 Routine and complex business situations and corresponding AI.

the business situations evolve from routine to complex and what is required in terms of corresponding support to those business situations. While AI provides active support by reengineering of processes, these increasingly complex situations also require best practices based on standards, quality, metrics, and training. Furthermore, CASE (Computer-Aided Software Engineering) tools can expedite the development and application of analytics to the complexity of business processes. As shown in Figure 1.5, the routine and well-defined data benefits by the application of known and routine ML algorithms in decision-making. This automation with help of ML can continue even with dynamically changing input data so long as the processes have not changed. When both data and algorithms change, the complexity of business processes is very high and a complete revamping of processes is required. This is the intelligent optimization of processes. Eventually, when the issues of biases, ethics, and human values come into picture, AI is not sufficient; this is where humanization is inevitable. The entire application of AI and its use in making decisions is an iterative and incremental (Kaizen-like) process.

Comparing automation and optimization

Table 1.1 compares automation with optimization across key parameters of digital business: data, devices, communications, analytics, and mind.

Figure 1.6 shows the conceptual difference between automation and optimization. An existing process is shown with two activities (A and B)

Table 1.1 Automation vs optimization activities of businesses corresponding to their parameters

	AI/ML application	
Parameters	*Business automation*	*Business optimization*
Data	Collect and store a large amount of data without worrying about where and how it can be strategically used.	Strategically consider how data provides an opportunity to change business processes using patterns. Macro- and alternative-data usage.
Devices (IoT)	IoT devices are used only in data collection without strategy. Sensors providing high-volume data.	Strategy for the collection of relevant data from devices, its processing (on the device, network, or Cloud).
Communications (networks)	Speed and volume of data over the network are key criteria for automation. Security is considered but not as a balancing act with performance.	Dynamically changing speeds – depending on the importance and relevance of data and analytics to the location and urgency of use.
Analytics (algorithms)	Mono-dimensional insights are used to improve existing processes.	Multidimensional insights are used to challenge the existing processes.
Mindset (people)	Let's do the same things faster.	Why are we doing and what we are doing?

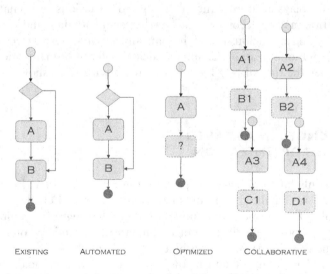

EXISTING   AUTOMATED   OPTIMIZED   COLLABORATIVE

Figure 1.6 Automation, optimization, and collaboration of business processes.

and a decision box. From the business process context, primary challenges center around gathering business requirements that can define the scope of change. Current business processes provide a point of reference for defining potential collaborative processes, associated partner users, and their business applications.

When AI is applied to automate the process, the decision box and the activities (A and B) still remain in the process. The value of automation is the increase in speed and accuracy of the process, but the process remains the same. Once the process is subjected to optimization, its entire flow is revised and reengineered. AI-based algorithms can potentially get rid of the decision box and also an activity (B, in Figure 1.6). With an optimized process, the business is able to reach out to other businesses much more easily than with only an automated process. This results in a suite of collaborative business processes, as shown in Figure 1.6 to the right. Optimizing a single business process provides a limited value. Figure 1.6 points to the need for holistic changes to the business. Further, substantial value is generated in a digitally interconnected world by collaborating with business partners. Automating and optimizing business processes across the enterprise require management of their disparity and heterogeneity. Collaboration and maintaining connectivity are crucial in promoting ROI in AI technologies and resources. Optimization leads to collaboration across the industry and across a collaborating group of businesses.

Intelligent humanization

Humanization introduces subjective human values in the decision mix. Subjectivity plays an important role in "good" decisions. Automation and optimization may be devoid of values as they deal with data and algorithms. Superimposing a judicious mix of human subjectivity on AI-based analytics is intelligent humanization. Chapter 10 deals with this humanization aspect of AI under the title of NI. The mixing of NI with AI is shown earlier in Figure 1.4.

CHALLENGES IN AI-BASED BUSINESS OPTIMIZATION

AI is certainly far from being the panacea it is touted by some. In fact, most discussions are balanced in terms of the advantages and challenges. In the context of BO, it is important to understand the challenges and limitations of AI. AI limitations, in particular, range from human dependency on machines, displacement of humans, irreversibility of AI, and the lack of empathy in machines.[45] Developing an understanding of these limitations holds the potential for successful application of AI in BO. The limitations of AI in terms

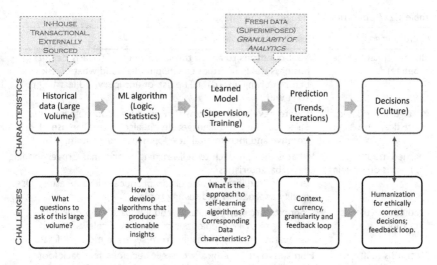

Figure 1.7 Application challenges corresponding to the characteristics of ML-based prediction model.

of application, business, culture, knowledge management, user experience, cybersecurity, and collaboration are discussed next.

Application challenges

Figure 1.7 summarizes the application challenges in ML-based prediction models. Each ML characteristic represents a challenge. Figure 1.7 also depicts the process of arriving at decisions from data.

Table 1.2 lists the characteristics of ML and the suggested approach to handling the challenge.

Business challenges

The most common challenges and issues related to AI-based BO are linked to four integral parts of collaboration: connection, communication, coordination, and commitment.[46] The complexity of the challenge usually varies according to the size, heterogeneity, and disparity of the enterprise's systems. Furthermore, the complexity also results from overlapping challenges, issues, or risks. Challenges, issues, and risks of embedding data-driven decision-making in business processes are discussed next.

Organizational culture challenges

The redesign of the business, mentioned earlier, also implies a major change in the sociocultural aspect of the business. Such change requires an

Table 1.2 Application challenges in the ML space

Characteristics	Descriptions and challenges
Historical data (large volume)	What questions to ask of this large volume? Clearly define business objectives that can help understand what to ask before delving into data. The use of unsupervised learning to start with.
ML algorithm (logic, statistics)	How to develop algorithms that produce actionable insights. Select relevant and manageable ML algorithms. The use of iterative and incremental development of algorithm.
Learned model (supervision, training)	What is the approach to self-learning algorithms? Corresponding data characteristics? Verify that the learned models do produce relevant results. Use quality approaches and testing tools.
Predictions (trends, iterations)	Context, currency, granularity, and feedback loop. Fine-tune learning to make predictions concrete and relevant. Enact comparative study of results from separate models of data.
Decisions (culture)	Humanization for ethically correct decisions; feedback loop. Verify that the predictions made by the ML algorithm are bias-free and ethically sound.

understanding of change management by the leadership. Changes in the organizational culture include changes to the roles, their reporting hierarchies, and the way they interact with the customers using data-driven decision-making. Shifting to data-driven organizational cultures requires the acceptance of technologies that disrupt the existing culture and norms. People may be reluctant to hand over routine processes to automation. This challenge requires a change in the mindset of the leaders, decision-makers, service providers, and customers. Furthermore, the rights, responsibilities, and accountabilities to the users and decision-makers also undergo change. This change is controlled by good IT and business governance. A set of principles and best practices across the organization help in handling the change that includes dataset, toolset, and mindset change. For example, providing users the training with the right tools, processes, and standards and mentoring the staff that may be reluctant to do so.

Adapting successful best practices and setting guiding principles for their use in data-driven decision-making are important elements of organizational culture.

Leaders play in important role in ensuring that the decision-making process is not automated to an extent where it has lost the ethical and moral values that are important to the users. The decision-making still needs to remain humanized. The cultural challenges of an organization using AI in decisions are further underscored by the uncertainty in measuring customer satisfaction as a result of BO. Customer satisfaction is a subjective element that also keeps changing rapidly depending on the context in which the customer finds herself. AI is not able to measure accurately this subjective element in customer satisfaction and, therefore, needs NI to be superimposed.

Knowledge management challenges

The entire AI-based decision-making generates new and interesting knowledge for the organization to use in an iterative and incremental manner. This requires the organization to include knowledge management (KM) as a part of its overall BO strategies. For example, KM needs categorization – organizing, cataloging, and indexing business application processing information based on user access patterns while improving user productivity. KM also needs repositories and data stores – collecting structured data, unstructured data, and metadata for current, interim, or future collaborative enterprise initiatives utilizing advanced technologies and techniques.

Process modeling and reengineering of processes are based on understanding the gaps between the current and future processes. BO strategies need to analyze and define these gaps, and outline the process of change management. KM helps in the integration of processes and AI with change management and helps in getting people using those changed business processes.

Visualization and reporting

An important yet oft lagging part of BO strategies is the visualization of AI-based analytics and their reporting. Visualization is integral to success in BO because it includes analysis and design of the reports from the analytics that are of value to the decision-makers. Visualization and reporting require an understanding of the specific needs of the users as well as the needs and nuances of presentation styles. Visualizations include, for example, creating on-the-fly or ad hoc reports of user metrics (e.g., security logging) to facilitate decision-making processes of managers and executives to improve the quality of collaboration. Visualization and reporting depend on the quality of data, analytics, and devices.

User experience challenges

Visualization leads to another important concept in the success of BO – that of overall user experience. This user experience (UX) includes not only the visuals but also their timings, relevance, and the performance of the overall solution. UX is subjective because it is primarily based on the perception of the user. Therefore, UX can shift for different users of the same process. In fact, UX can also shift for the same process and the same user at different times. Offering succinct experience and fulfillment to all users is an important challenge as BO is implemented.

Some important elements of UX include ease of access for data and business applications, ease of navigation of relevant information (using remote or on-site collaboration facilities for information exchange) from different user access devices (PDA, wireless, or browsers) and user comfort levels (from home or on the road).

Additionally, presentation can include a multitude of configurable user interfaces that allow the users to customize the visualization to suit their own needs and context. Enabling the users to utilize the available technologies to their fullest and establish collaborations to provide them with newer and wider choices is the key to UX.

Quality of service (QoS) is the metric that impacts UX. For example, performance and reliability (two important QoS metrics) play a crucial role in UX and, at times, even greater than the actual functionalities offered by the analytics.

UX is based on continuous improvement of offerings based on the needs and context of the users and a smart feedback loop.

Cybersecurity challenges

Cybersecurity is of utmost importance in BO, and it is discussed in detail in Chapter 9. Cybersecurity primarily deals with the sanctity of data across its entire usage. Security also ensures physical security of devices, privacy, and confidentiality of data. There are a number of techniques used in ensuring the security of optimized business processes: for example, multifactor authentications using handheld devices. Following are some additional cybersecurity considerations:

- Single sign-on capability – Formalizing access to appropriate business applications or services based on policies, procedures, agreements, or profiles.
- Identity management – Verifying the identity of users by matching or validating encrypted passwords, digital certificates, and public or private keys for participation in collaborative enterprise initiatives.
- Access authorization and validation – Defining special privileges for users to access specific information to perform certain activities with or without updating any information or to control specific tasks.
- Logging user metrics – Maintaining a log of user access (usage, time, and a number of business applications or services accessed) to determine access violation, denial, and breach of policies or for audit trails.
- User account management – Creating, managing, and supporting user accounts and access privileges; supporting activities such as forming user groups or communities of practice.

Collaboration challenges

A crucial opportunity with the digitization of business processes is their extended ability to reach outside the organizational firewall to business processes in other partnering businesses. Data sharing and information exchange across an integrated and extended enterprise lead to further exchanges across other enterprises. This sharing of data through application programming interfaces (APIs) extends the reach of a business and enables it to offer wider choices of products and services to its customers. Following are some important, additional considerations for collaborations in an AI-enabled BO effort:

- Unified customer view by sharing the customer profiles across businesses, eliminating redundancies in data and processes, and providing the customer with a single point of contact (SPOC).
- Managing messages and events across businesses by delivering alerts, messages, updates, or notifications for the entire group of collaborating businesses.
- Establishing and promoting user communities that would share experiences, discuss challenges, and exchange experiences across multiple users with common goals or interests (partners, vendors, clients, or associates).
- Sharing responsibilities and accountabilities of providing value to the customer through coordination and monitoring of various activities and decision-making as appropriate; commitment to delivering the right information at the right time (maintaining accountabilities and responsibilities).
- Establishing and managing governance, risk, and control across multiple businesses by the application of collaborative enterprise architectures (Eas), and involving business leaders and staff across these partnering enterprises.

COVID-19 pandemic and digital business

The COVID-19 pandemic was a historic turning point in the world – upending all earlier business values, disrupting the conventional business operations, challenging the accepted business wisdom, and throwing the digital world into the limelight. COVID-19 generated an interest in technologies beyond analytics. For example, the pandemic moved almost the entire world to online work. Working From Home (WFH) became a norm. The pandemic established a new business order based on the digital world. Furthermore, online sales, purchases, deliveries, and services have increased multifold, demanding a deep rethink of customer value. AI impacts all these business functions during optimization.

The world of DS has a role to play during and post-pandemic. The interest and importance of application of AI in understanding businesses and the future trends and patterns is vital during the pandemic. Data-driven decisions enable the management of spaces and people in an optimum way during the WFH era. AI can improve efficiency and lower risks in the new business world order. This is so because AI can be used to analyze and identify problems and challenges before they occur.

The WFH churns out phenomenally high quantities of data with high velocity. Human capabilities fall short of understanding this Big Data. This is where AI helps businesses understand "what is going on out there?" in an ongoing manner.

Communications are crucial in this shift to online world. Devices (especially IoT – Internet of Thing) accompany communications as an important factor in the successful shift to WFH. The pandemic provides the impetus for adaptation and adoption of the digital "mindset." AI creates opportunities to understand the style, throughput, and challenges of the workforce based on data. For example, less office space and more detailed and structured online working require data to justify these styles of working.

CONSOLIDATION WORKSHOP

1. What is AI and how it is related to BO?
2. How do AI, ML, and Big Data intersect? Give an example of each intersection.
3. Describe the paradigm shift in terms of the culture that a BO initiative leads to.
4. What are automation and optimization? How are the two different?
5. Why is collaboration across the industry by multiple businesses important in order to derive full benefits of BO?
6. Outline the challenges in BO with AI.
7. Why can businesses not put themselves entirely in the hands of AI? What are some of the limitations of current AI technology?
8. What is the role of NI and humanization in BO?
9. What are the different ML types? Describe with examples.
10. What is the importance of supervised learning in BO?
11. What are the key characteristics of unsupervised learning? How does it relate to DL?
12. What is RL and how can it be used in BO?
13. Discuss the increasing complexity of business processes and the role of AI in them?
14. How would businesses handle the COVID-19 pandemic without AI and automation?
15. What concrete lessons about business practices have enterprises learned from the impact of COVID-19 disruption?

NOTES

1. Sivarajah, Uthayasankar, Muhammad Mustafa Kamal, Zahir Irani, and Vishanth Weerakkody. 2017. 'Critical analysis of Big Data challenges and analytical methods', *Journal of Business Research*, 70: 263–86.
2. Davenport, Thomas H. 1998. 'Putting the enterprise into the enterprise system', *Harvard Business Review*, 76: 121–31.
3. Unhelkar, B. 2018. *Big Data Strategies for Agile Business*. (CRC Press, Taylor & Francis Group/an Auerbach Book, Boca Raton, FL). Authored ISBN: 978-1-498-72438-8 (Hardback), Foreword Prof. James Curran, USFSM, Florida, USA.
4. Finlay, Steven. *Artificial Intelligence and Machine Learning for Business: A No-Nonsense Guide to Data Driven Technologies*. Relativistic, Great Britain 2018.
5. Miller, T. 2019, February. 'Explanation in artificial intelligence: Insights from the social sciences', *Artificial Intelligence*, 267: 1–38.
6. See note 4. IBID for reference #4
7. See note 4.
8. Ada Lovelace. Augusta Ada King, Countess of Lovelace was an English mathematician and writer, chiefly known for her work on Charles Babbage's proposed mechanical general-purpose computer, the Analytical Engine; https://en.wikipedia.org/wiki/Ada_Lovelace accessed 30 Sep, 2020.
9. Tiwary, A., and Unhelkar, B. 2018. *Outcome Driven Business Architecture*. (CRC Press, Taylor & Francis Group/an Auerbach Book, Boca Raton, FL), Co-Authored.
10. Agarwal, A., and Unhelkar, B. 2016. 'Context driven optimal granularity level (OGL) in Big Data analytics', In *Proceedings of Society for Design and Process Science Conference*, SDPS2016, 4–6 Dec, 2016, Orlando, FL.
11. Chen, X., and X. Lin. 2014. 'Big Data deep learning: Challenges and perspectives', *IEEE Access*, 2: 514–25
12. Zhang, Qingchen, Laurence T. Yang, Zhikui Chen, and Peng Li. 2018. 'A survey on deep learning for Big Data', *Information Fusion*, 42: 146–57.
13. See note 9.
14. See note 9.
15. See note 3.
16. Cabena, Peter, Pablo Hadjinian, Rolf Stadler, Jaap Verhees, and Alessandro Zanasi. 1998. *Discovering Data Mining: From Concept to Implementation*. (Prentice-Hall, Inc., Upper Saddle River, NJ).
17. Schmidt, Cecil, and Wenying Nan Sun. 2018. 'Synthesizing agile and knowledge discovery: Case study results', *Journal of Computer Information Systems*, 58: 142–50.
18. Grossi, Valerio, Andrea Romei, and Franco Turini. 2017. 'Survey on using constraints in data mining', *Data Mining and Knowledge Discovery*, 31: 424–64.
19. L'Heureux, A., K. Grolinger, H. F. Elyamany, and M. A. M. Capretz. 2017. 'Machine learning with Big Data: Challenges and approaches', *IEEE Access*, 5: 7776–97.
20. Ayodele, Taiwo Oladipupo. 2010. 'Introduction to machine learning.' *New Advances in Machine Learning (IntechOpen)*. Carbonell, Jaime G., Ryszard S. Michalski, and Tom M. Mitchell. 1983. 'Machine Learning Part I: A Historical and Methodological Analysis.' In *AI Magazine*. Birmingham, AL, USA

21. Unhelkar, B., and Gonsalves, T. 'Enhancing Artificial Intelligence decision making frameworks to support leadership during business disruptions', " In IT Professional, vol. 22, no. 6, pp. 59-66, 1 Nov.-Dec. 2020, doi: 10.1109/MITP.2020.3031312.

22. Thompson, Wayne. Free report downloaded on 17th Nov 2016; based on presentation given at the Analytics 2014 Conference, Manager of Data Sciences Technologies at SAS http://www.sas.com/en_us/offers/sem/statistics-machine-learning-at-scale-variant-107284/download.html#.

23. Madani, Badis, Hosam Alagi, Björn Hein, and Aurilla Aurelie Bechina Arntzen. 2017. 'Machine learning for capacitive proximity sensor data.' In Society for Process and Design. Madani, Badis, Hosam Alagi, Björn Hein, and Aurilla Aurelie Bechina Arntzen. 2017. 'Machine learning for capacitive proximity sensor data.' In Society for Process and Design.

24. Shalev-Shwartz, Shai, and Shai Ben-David. 2014. Understanding Machine Learning: From Theory to Algorithms. (Cambridge University Press).

25. Rouse, M. 2011. 'Machine learning definition'. http://whatis.techtarget.com/definition/machine-learning.

26. See note 24

27. Provost, F. and Fawcett, T. Data Science for Business: What You Need to Know About Data Mining and Data-Analytic Thinking. (O'Reilly Media, Inc). 2013 – revised 2019. Sebastopol, California

28. Weng, Sung-Shun, Ben-Jeng Wang, Ruey-Kei Chiu, and Sheng-Hung Su. 2006. 'The study and verification of mathematical modeling for customer purchasing behavior', Journal of Computer Information Systems, 47: 46–57.

29. Da Cunha, Catherine, Bruno Agard, and Andrew Kusiak. 2006. 'Data mining for improvement of product quality', International Journal of Production Research, 44: 4027–41.

30. Havenstein, Heather. 2006. 'IT efforts to help determine election successes, failures: Dems deploy data tools; GOP expands microtargeting use', Computerworld, 40: 1.

31. Interesting discussions in: Baesens, B., Daniel, Roesch, and Harold Scheule. Wiley 2016. Credit Risk Analytics: Measurement Techniques, Applications, and Examples in SAS, ISBN: 978-1-119-14398-7.

32. Fair Isaac Corporation (FICO) Scores provide an industry-standard for scoring fair creditworthiness for lenders and consumers.

33. Robert, Christian. 2014. Machine Learning, A Probabilistic Perspective. (Taylor & Francis).

34. Witten, Ian H., Eibe Frank, Mark A. Hall, and Christopher J. Pal. 2016. Data Mining: Practical Machine Learning Tools and Techniques. (Morgan Kaufmann Burlington, Massachusetts).

35. See note 27

36. Enke, David, and Suraphan Thawornwong. 2005. 'The use of data mining and neural networks for forecasting stock market returns', Expert Systems with Applications, 29: 927–40.

37. Wei, Chih-Ping, and I-Tang Chiu. 2002. 'Turning telecommunications call details to churn prediction: A data mining approach', Expert Systems with Applications, 23: 103–12.

38. Müller, Andreas C., and Sarah Guido. 2016. *Introduction to Machine Learning with Python: a Guide for Data Scientists*. (O'Reilly Media, Inc Sebastopol, California).

39. Brachman, Ronald J, Tom Khabaza, Willi Kloesgen, Gregory Piatetsky-Shapiro, and Evangelos Simoudis. 1996. 'Mining business databases', *Communications of the ACM*, 39: 42–48.

40. See note 11.

41. See note 12.

42. Gonsalves, T. 2017. *Artificial Intelligence: A Non-Technical Introduction*. (Sophia University Press, Tokyo).

43. Hazra, T., and Unhelkar, B., 0. *Enterprise Architecture for Digital Business: Integrated Transformation Strategies*. (CRC Press, Taylor & Francis Group Boca Raton, Florida).

44. Gonsalves, T., and Unhelkar, B. 2020, June 'Superimposing natural intelligence (NI) on artificial intelligence (AI) for optimized value in business decisions', *Cutter Business Technology Journal*, Ed. Murugesan, S., 33(6): 26–32.

45. Brown, B. *Artificial Intelligence and Machine Learning*, © Benjamin Brown Vancouver, British Columbia, Canada, Ch 9, 113–115.

46. Hazra, T. K. 2009, November. 'EA metrics deliver business value: Going beyond the boundaries of the EA program', *Cutter IT Journal*, 22(11).

Chapter 2

Data to decisions

Evolving interrelationships

THINK DATA

Data is a vast topic. Data encompasses observations, acquisition, recording storage, cleansing, analytics, security, and disposal. In the digital world, a simple query, a product search, a purchase, or a post-sale service request generates data. This data is typically stored on vast cloud servers. Data is then analyzed to produce insights that form the basis for business decisions. Understanding how data evolves into decisions is important for business optimization (BO).

The explosion in data globally is set to reach close to 150ZB by 2024.[1] The growth in data volume is fueled by growth in velocity (e.g., 5G), variety (e.g., unstructured, audio, graphics), and the need for veracity (e.g., quality, darkness). Business data is generated by internal business processes, staff, external customers, business partners, regulatory agencies, and user communities. User communities also generate data whose harvesting can lead to many interesting business ideas. Business data is typically stored on the Cloud. The challenge for business optimization is to sort the relevant data and analytics from the noise across multiple Cloud sources and ensure a well-understood evolution from data to decisions.

Data architecture based on EA[2] and OBDA[3] is required to integrate technology, storage, computing, and device requirements. Multiple data feeds using API and scalable processing power on the Cloud are essential. Data analysts analyze datasets and systematically look for the insights within them.

Think data: Handset, dataset, toolset, mindset

Data science establishes correlations between suites of data. Data science, however, is not able to (nor is it meant to) explain the underlying cause for patterns and trends. Correlation, not causality, is the theme of data science. Data analysis "plays" with the data, creates "what-if" scenarios, and supports business decisions. Such data analysis is subject to influences of changes in the input, frequency, and context of the data. Thinking about data requires an understanding of the myriad ways in which data can be sourced, analyzed, and retired. The tools and technologies and the creativity required to make use of data are part of data science. Furthermore, variations of data types, such as alternative data, may not even be directly related to the purpose of the analysis. An example case is the correlation between the satellite images of various car parks in Wuhan region and the possibility that COVID-19 started much earlier than declared by the WHO.[4] The challenge in the use of such data is its validity and currency – the factors discussed later in this chapter in using alternative data.

Figure 2.1 summarizes the four key strategic aspects of data applicable to BO: the handset, dataset, toolset, and mindset. Keeping all four aspects in *Think Data* in balance is important in BO.

- Handset – this aspect of *Think Data* includes smartphones, IoT devices, and sensors. *Think data* starts with this important and mainly automated source of data. Users continuously interact with the business through their handsets. The volume and velocity of data depend on the use of the handset. Sensors, although not held in hand, automatically generate a large-volume and high-velocity data.
- Dataset – this aspect of "thinking data" deals with the storage security and utilization of data. The data characteristics of volume, velocity, variety, and veracity are applied in managing datasets. For example, structured sensor data needs SQL storages and unstructured blogs or smartphone data needs NoSQL datasets. No backups or archives are possible with Big Data, especially alternative data. Data-driven digital business is not limited to using its own data. While costs, prices, and transaction volumes are examples of data generated by the business,

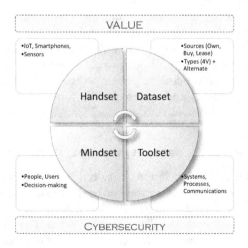

Figure 2.1 Think data – handset, dataset, toolset, mindset.

there are meta-data and alternative data generated by interactions among third parties and customer sentiments.

- Toolset – includes the tools for communication, storage, and processing. For example, analytics in the Cloud, analytics-as-a-service, visualization, and the various aspects of embedding analytics in business processes are thought through here. ERP and CRM solutions are also part of toolsets. Tools used in providing cybersecurity analytics are also included.
- Mindset – this is the user-centric view of data usage. The mindset also includes strategies for dealing with the inevitable change when data analytics is embedded in decision-making. Handling mindset requires interdisciplinary teams comprising data scientists, domain experts, business process experts, finance and economics professionals, and human resource (HR) managers. Training of staff and educating users is a part of "Mindset" in optimizing business processes. Mindset is an indication of user sentiments that impacts predictions.

As also summarized in Figure 2.1, underpinning *Think Data* is "Think Cybersecurity." Each aspect of the "think" data needs to consider cybersecurity. For example, handset needs physical security, dataset needs secured cloud storage, toolset includes encryption, and mindset requires good user habits and awareness.

The over umbrella of *think data* is "Value." Therefore, all four aspects of *think data* need to be in balance.

Various aspects of think data

Understanding data implies understanding its various dimensions from sourcing and storage through the security and retirement. The entire gamut of application of the keyword "data" is summarized in Table 2.1.

Table 2.1 provides a holistic checklist of data as applicable in business optimization. This corresponding thought focusing on each aspect of data is also provided in Table 2.1.

Data characteristics

Strategic thinking data for business optimization includes decision-making that combines analytics (explicit) and human (tacit) thinking. The entire decision-making process is iterative and incremental resulting in business agility.[5]

Embedding data analytics in business processes makes them more efficient and effective. Customer experience is also personalized as a result.

Data is not limited to structured sets. Low-volume data and unstructured data with variety has the potential to provide valuable insights. The NO

Table 2.1 Think data for business optimization

Data	Think! (for business optimization)
Data science	Think overall principles, concepts, and strategies for the application of data to business. Datasets, handsets, toolsets, and mindsets are included in Data Science.
Data analytics	Think algorithms to process the data and provide results. Descriptive, predictive, and prescriptive analytics together with supervised, unsupervised, and reinforced learning techniques are examples of analytics.
Data architecture	Think of technical alignment.
Data bases	Think SQL and NoSQL databases storing data locally, on the Cloud, and on user devices in both structured and unstructured formats.
Data communications	Think networks and transmission infrastructure including its speed, security, and context. Wired, Wi-Fi, and Cellular for data and analytics (especially as they are offered as services).
Data context	Think of the business context in which data will be used. The urgency, importance, and relevance of the business process provide an additional context.
Data costs	Think costs associated with per unit (e.g., MB or GB) of data (e.g., storage, backup, and mirroring costs). Costs of sourcing data from vendors are included. These "cost"-related thoughts balance the expenses and risks of BO with returns (ROI).
Data density	Think of what reality the data represents. The more compact a data point in terms of representing the reality, the denser the data. And, the denser the data, the more effort and tools are required.
Data manipulation	Think location of data and data movements. Manipulation is associated with data retrieval, preparation, and processing.
Data mining	Think layers and depths of datasets; think the nature of data – static vs. dynamic. The algorithms for mining depend on the understanding of data layers and architecture.
Data presentation	Think visualizations, users, IoT devices, screen real estate, and the relevance of presentation to the user in making decisions; video, audio, and sensor (buzzers) are examples of presentation types.
Data privacy	Think legal aspects of data and repercussions of breaches on businesses. These privacy aspects change depending on the source, ownership, usage, and retirement of data.
Data processes	Think business processes that use data analysis. Process optimization is based on an understanding of data usage within the business process.
Data quality	Think verification and validation of data, processing, and visuals. Think of the veracity of data, which is an ongoing effort based on continuous testing. Think also of the amount and type of data to be used for this testing versus the one to be used for training the AI algorithms. Is the business generating dark data? Is there dark data in the environment of the collaborating partners?
Data security	Think cybersecurity in terms of all aspects of data – including and especially during communications via networks and processing. User behavior forms the "soft" Should be soft aspect of security.

(Continued)

Table 2.1 (Continued) Think data for business optimization

Data	Think! (for business optimization)
Data sources	Traditional, meta, and alternative data sources. Owned and leased data and also data scraped from sites. Synthetic data is generated internally to test and also "pad" datasets for analytics.
Data statistics	Think clustering and classification of data and the use of relevant statistical techniques to make sense of the data. Training and testing data are essential for statistical analysis.
Data tools and technologies	Think current and future tools, and technologies related to sourcing, storage, analytics, and application of data.
Data usage	Think the lifecycle of data – from observation to retirement and everything in between. Vast storage of data with no usage is a burden to the business.
Data value	Think end goals of customer satisfaction for a business. The entire BO effort is directed toward customer value, and data should help a business reach that.
Data warehouse	Think storage of transactional and large historical data and associated systems for retrieval and analytics. Think strategies for mining historical data with descriptive analytics.

SQL databases[6] provide opportunities for analytics and influence business decision-making even without high volume. ML handles not only structured data, but also semi- or unstructured data irrespective of its volume.

This data also has rich meta-data (e.g., parameters around a data point) associated with it. This meta-data provides aggregated information for business – such as potential demand for a product or service from a group of customers. Continuously growing data holds the promise for continually improving insights.

The data, its context, and meta-data form a "story." These data stories help understand customer behavior, spot market trends, and satisfy compliance needs. Analytics enable a business to proactively respond to changing external and internal situations.

Data – especially large volume, high velocity, and unstructured data – can appear to be chaotic. Data strategies find a time-rhythm and a structure-pattern in that data. These rhythms and patterns are used to predict the interest rates tomorrow, the weather next week, and the best airfare a month in advance. Data science identifies trends and patterns through exploration, matching, referencing, correlating, and extending data and meta-data.

Raw data evolves and transforms into information, analytics, and knowledge. Knowledge drives decisions related to products and services.

Classifications and categorization are the starting point for the application of statistical techniques to date. ML creates and uses models based on these techniques to generate insights.

Data is extracted for the repository of raw data for training and testing purposes. Features are defined in the training data. Accurate models of data and processes are used for further training and testing of models.

Data as enabler of optimization

Data through the range of *think data* is an enabler of process optimization. Detection, predictions, and prescriptions are embedded in the business functions. Identifying the right data type and having quick access are important for integrating within the decision-making process. Testing the veracity of the dataset is required on a continuous basis. An integrated analytics platform promotes idea-sharing and generates greater efficiency. Combining this with traditional data can lead to differentiated insights.

Following are some examples of how *think data* enables optimization:

- Location and density data of mobile devices predict economic outlook as it indicates increased human activity. This prediction enables the organization to develop strategies for newer products and services.
- Analyzing prices of millions of online products enables an understanding of shifts in global demands and, therefore, on pricing and marketing.
- Identifying changes to purchasing habits of customers to cross-sell and up-sell new products and services.
- Correlating credit card transactions, geolocation, and app downloads provide insights into how the business is viewed by the customers and which other businesses it should collaborate with.
- Correlating search engine data with social media data enables predicting earnings.
- Twitter and other social media feeds together with sentiment analytics enable the capabilities to direct resources.

DATA TO DECISIONS PYRAMID

Data evolves through five layers to become decisions. As shown in Figure 2.2, the evolution of data to decisions is a systematic, incremental process that impacts and is impacted by artificial intelligence (AI). This evolution requires a detailed understanding of handset, dataset, toolset, and mindset. Figure 2.2 starts with observations which, when recorded, become data. Observations are the start of data collection. Data goes through a process to become information of analytics in order to provide knowledge and insights. Analytics achieve this by correlating wide-ranging and dispersed suites of data. This is made possible due to the shareability of Big Data on the Cloud. Analytics support predictions and prescriptive advice as the data evolution continues.

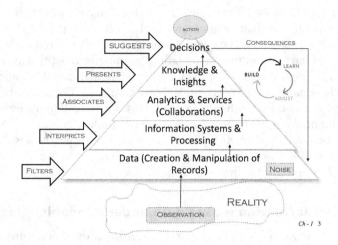

Figure 2.2 The pyramid of evolution of data to decisions and the AI impacts.

Decision-makers utilize these insights to make decisions. Good decisions consider all the layers of data evolution and their contribution to the final outcomes. Decisions are put into effect by action. Action generates consequences and further observations, and the cycle continues.

Observations are of the reality. Observations are subjective as they are influenced by the perceptions of the observer. Observations can be a business transaction, customer interaction, or a piece of equipment. Facts are observed in an unbiased manner and recorded as data. In addition to humans, machines and sensors (e.g., IoT) also record observations. Multiple observations of the same reality over time, place, and people provide increasing confidence in its accuracy.

Observation and decisions are primarily subjective (tacit). Data, information, analytics, and codified knowledge are mainly objective (explicit).[7] Only the objective aspect of decision-making can be automated. Intelligence in automation and optimization is a balancing act between the tacit and the explicit aspect of business decision-making.

Below is a brief description of the five layers of the data to the decision pyramid of Figure 2.2.

Layer 1: Data is a record of observations

Data is a suite of observations consolidated and organized in an objective manner. This layer represents the creation and manipulation of records. Storages can vary vastly, from the neatly organized rows and columns of a massive columnar database, through to the complex, multimedia, data warehouses containing audio, video, photos, and charts. Data is objective, storable, shareable, and subject to varied analysis. The quality of data is enhanced through filters. Data is quantitative and qualitative in nature. Big Data is characterized by high volume, velocity, and variety. Data contains noise that has to be reduced.

Sharing of data across the organizations eliminates repetitive and redundant data. For example, a customer demographic data, such as name and address, is usually stored by the telephone company. Therefore, this data need not be stored by the bank. Instead, this data is collaboratively available to the bank from the telephone company under "contracts." Sharing data through well-connected, reliable, and trustworthy partners is the basic form of collaboration among organizations. Data sharing is usually over the Cloud in order to facilitate collaborations.

Layer 2: Information makes data understandable

Level 2 represents the processing of data to create information. Information is the systematic use of data in business processes. Data, on its own, is not meaningful, whereas information based on the data provides meaning. Big Data is both a challenge and an opportunity in the quest for BO. For example, bank data is processed to generate information on demographic behavior patterns, such as spending styles, income groups, and geographical nuances of the customer. Creating information and understanding the decision-making process are important Level 2 activities.

Processing vast and complete datasets reveal hidden semantics. Process is activities and steps undertaken in businesses, like that of opening an account in a bank or withdrawing cash from an ATM. While minor variations in each of these processes are accepted, the fundamental process remains the same. By creating a basic process model for opening an account, it is possible to embed analytics in it. Rules and regulations can be commonly applied to these information system processes. Furthermore, agencies specializing in fraud detection can inspect the information for suspicious transactions.

Layer 3: Analytics and services (collaborations)

Analytics establish correlations between data. Analytics identify patterns and trends within those data. The more the data, the better is the output. Also, data analytics is not limited to analytics on a singular type of data. Data is sourced with the help of services from multiple and widely varied databases (typically on the Cloud) and a relationship established between them in order to perform data analytics. Analytics itself is offered as a Service[8] on the Cloud. Electronically established collaborations among other partnering businesses enable offering of a wide spectrum of services to the clients of the organization. Therefore, analytics and services together ensure data and information have evolved to a highly sophisticated level of usage. Personalization on a highly scalable basis results from collaborative services.

Layer 4: Knowledge and insights

Knowledge can be understood as rationalization and correlation of information through reflection, learning, and logical reasoning. The knowledge discovery processes are categorized under six distinct groups (Helmy, Arntzen Bechina, and Siqveland 2018)[9,10]: Formulating the Domain Application, Data Acquisition, Data Preparation, Machine Learning (ML), Evaluation and Knowledge Discovery, and Deployment. The focus of this 4th layer in the pyramid of Figure 2.2 is to correlate information and use analytics to create new knowledge.

Information can lay in silos, mainly dictated by the original classification of data upon which it is based. Converting this information to knowledge is a multilayered process. Analytics combined with information produces knowledge. Big Data provides a greater opportunity to correlate between otherwise separate islands of information within and outside an organization as compared to normal data. This is so because almost all Big Data is stored in the cloud.

The level also shares knowledge about an individual or a group of customers or users across multiple organizations. For example, correlation between the information about a customer (person) available to one organization and other bits of information (such as geographical location or spending habits) available to another organization can be established. These wide-ranging correlations produce new and unique knowledge about that customer that is not possible with analytics in a single organization. Knowledge can extend the predictability of behavior beyond one customer to an entire customer group.

Layer 5: Decisions

Decisions are made up of explicit inferences based on insights and extensive correlations among widely dispersed knowledge and tacit mindset of decision-makers. While data, information, analytics, and (to a large extent) knowledge are considered objective, decisions are a human subjective trait. Decision makes the use of tacit human factors such as personal experience, value system, time and location of decision-making, sociocultural environment, and ability to make estimates and take risks. The implicit form of knowledge of the user is often cognitive "intuition" about how the input, the process, and the output are related.

AI suggests actionable knowledge in order to support decisions. Strategic use of Big Data and AI leads to insightful decisions. Analytical outputs are combined with the individual decision-maker's ability to consistently distinguish the importance, relevance, context, and organizational principles in decisions. The decision-maker comprehends and balances AI with NI for decision-making. Decision-making is an agile, iterative process with the consequences of the decisions providing fresh data as input for the next discussion.

Intelligence is not something that can be organized and placed in databases and transferred to others through training. Decisions lead to actions that are fully mature implementation of the data decision pyramid aiming at the value goal. Enhancing the customer experience in the most effective and efficient way is the purpose of data to decisions. Customer groups can also collaborate and support decisions to achieve higher value for themselves. Collaborations across organizations require an exchange of data and information across highly porous electronic boundaries.

BIG DATA TYPES AND THEIR CHARACTERISTICS FOR ANALYTICS

The 3+1+<u>1</u> (5) Vs of Big Data

BO uses Big Data in its analytics. Understanding the characteristics and nature of Big Data improves the opportunity of its usage in decisions. These characteristics are summarized in Figure 2.3. Big Data is initially characterized by high *volumes*. *Velocity* and *variety* emerged as important additional characteristics of Big Data. Processing capabilities and storage are required to be enhanced as Big Data moves beyond the confines of a traditional data warehouse. *Veracity* relates to the quality of data. Business optimization approaches the fifth characteristic of "V"alue in a strategic manner.[11] BO comprehends the new possibilities for business with these characteristics. *Value* in this mix of characteristics is the most important yet hidden characteristic of data. The context of data and the concept of Agility in business are important contributors in extracting the value that is hidden within the data. As new solutions continue to emerge, users need to quickly understand the implications of the new forms of decision-making.

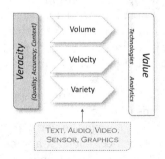

Figure 2.3 Detailed characteristics of Big Data's 3 + 1 + 1 Vs and the types and categories of data.

Large, historical static datasets are captured, communicated, aggregated, stored, and analyzed. But data isn't static.[12] Instead, it is a combination of both static (large volume) data and dynamic data (high-velocity) data such as being streamed from IoT devices.

Big Data is defined as high-volume, high-velocity, and/or high-variety information assets that require new forms of processing to enable enhanced decision-making, insight discovery, and process optimization.[13] Large bodies of data also have encrypted patterns that represent knowledge[14] preserved from the past and providing invaluable hints, tips, and even concrete solutions to challenges being experienced in the present.

While volume, velocity, and variety are inherent to Big Data, veracity and value require a business-oriented and strategic approach to handling data. Data, often associated with text and numbers, can take the shape of an audio or video file in the Big Data space. Big Data deals with large volumes (e.g., ticker of a share market) and a variety of sources (e.g., from a website, a blog, or an IoT device). Text itself can be both structured and unstructured. For example, a form filled out online will have a structure associated with the fields. When such forms run into millions and the data entered within those forms needs to be analyzed, the structured data starts moving into the realms of Big Data. On the other hand, lesser volume but highly unstructured data (e.g., descriptive customer feedback or a blog) is also in the realm of Big Data. Data velocity is further related to clickstream, video streaming, and machine sensors. Audio-video are unstructured data requiring a translation into a structure before they can be analyzed. Such conversion of data to a structure for analysis is an iterative process.

Understanding data usage provides the basis for developing data-driven strategies for BO. The basic data usage pattern starts with observations. Once observations become data, that data goes through generate, record, store, secure, clean, retrieve, share, and use processes. Finally, good data usage is also concerned with the appropriate retirement of data. Four distinct yet interrelated aspects of the data usage pattern are summarized in Figure 2.4:

- Source and store data – This is the entry of data in the organization through direct, mobile, audio, text, etc.; volume, cloud, scalable, secured. Data analytic strategies are concerned with the sourcing of current and future data, its cost and how current, relevant and reliable the source and storage facilities are.
- Mine data– (analytics, patterns, collaborations, costs associated with data and mining; quality – cleansing). Data analytic strategies consider the relevance and the overall costs of mining data. Tools and technologies for data mining are considered here.
- Utilize data – (decision-making; granularity; feedback mechanism). Data analytic strategies include business process re-engineering in order to embed analytics in the decision-making process. The level of granularity is considered.

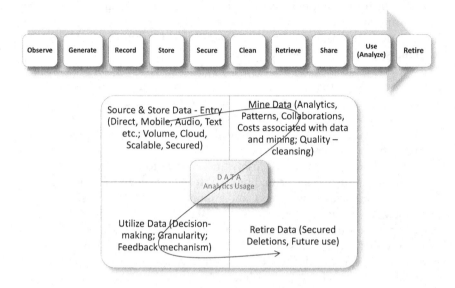

Figure 2.4 Data analytics usage pattern.

- Retire data – (secured deletions, future use). Data analytic strategies to deal with data that has already served its purpose. The usage of data requires an equally careful approach to its retirement and eventual deletion.

SOURCING OF DATA

Think Data and the data analytic usage pattern mention the sourcing of data. Substantial data is generated by machine sensors and IoT[15] devices as a source. The volume and velocity of such data are very high because it is generated without human intervention. The number of active IoT devices was estimated to be 21 billion by 2020[16] generating 40% of all data. These data through channels of interconnected sensors and devices are now available in massive quantities and in both structured and unstructured formats. The addition to this data is on a continuous basis with high velocity. Third parties and governmental bodies provide yet another source of data. Users generate and share content typically on the Cloud. Data analytics on the Cloud provides real-time insights that enable the optimization of processes. The end result is a vast collection of data from multiple sources, growing at a high rate and comprising varieties that have a potential value hidden in them.

Enterprise applications (CRM, ERP, SCM, and others) store this data. The volume, velocity, and variety of this data has bearing on its analytics.

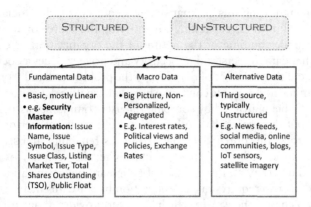

Fundamental Data	Macro Data	Alternative Data
• Basic, mostly Linear • e.g. **Security Master Information:** Issue Name, Issue Symbol, Issue Type, Issue Class, Listing Market Tier, Total Shares Outstanding (TSO), Public Float	• Big Picture, Non-Personalized, Aggregated • E.g. Interest rates, Political views and Policies, Exchange Rates	• Third source, typically Unstructured • E.g. News feeds, social media, online communities, blogs, IoT sensors, satellite imagery

Figure 2.5 Big Data categories based on sources.

Timely and accurate analytics are required in order to help optimize business. Analytics impact business decisions.

Figure 2.5 shows the additional categorization of Big Data in fundamental, macro, and alternative data.

- Fundamental – Basic data generated by the organization, mostly linear (e.g., master data on people, company data, share issue and type, listing exchanges)
- Macro – Big picture, non-personalized, aggregated (e.g., interest rates, political views and policies, exchange rates). Macro data is based on aggregated, fundamental data but it does not identify individual records.
- Alternative – Third source, typically unstructured (e.g., news feeds, social media, online communities, blogs, IoT sensors, satellite imagery)

Alternative data

Alternative data is not the organized transactional datasets. Alternative data is derived from third parties: for example, social media feeds, blogs, forums, and crowd-sourcing platforms. IoT devices are major sources of alternative data. Alternative data is a subset of Big Data, often unstructured and unidentifiable. Alternative data is sourced from outside the boundaries and controls of the organization and yet it references the organization. For example, a trending news item about the organization and the ensuing discussions by customers and other observers on the news item can be analyzed to find trends and patterns. The confidence levels of predictions improve with alternative data. Alternative data can provide validation for the main data and improve the granularity of predictions.

Analyzing and using this alternative data require AI technologies and corresponding and business capabilities. Identifying the source of alternate data requires an understanding of the nature of the organization or community providing that data, understanding security and privacy issues associated with the source, and the relevance of that data to the problem at hand. For example, analyzing an insurance quote can make use of alternative data about the property or vehicle being insured by sourcing the user community associated with that item – but the analyst has to be mindful of which Tweeter or Facebook community they are sourcing the data from, its reliability and security, and how well it has served the purpose of improving the accuracy of the insurance quote. Furthermore, the mindset of people in the organization who will utilize the data also requires due consideration because this niche data and its analytics can change the way in which an agent provides an insurance quote. Sourcing, analyzing, and integrating this alternative data in decision-making require more than AI tools. Expertise in the business domain has to couple with technical expertise in the mindset of the data science professionals.

While structured, transactional data is analyzed quantitatively, alternative data needs more creative approaches for sourcing and analytics. These approaches include analyzing the free-formatted blogs, opinions, likes and dislikes, and the speed with which a particular community is expressing its opinions. For example, if a car owner community posts a certain number of "likes" and favorable opinions for a new brand of car model within, say, 1 week, then it has a higher density of positive opinion than the same positive expressions over 4 weeks.

Alternative data potentially offers an advantage over only using traditional data. Opportunity to derive value from previously ignored and/or emerging datasets, often termed the "exhaust" of other business or communities, is the new edge in data analytics.

For example, consider the prophecy project still incomplete. prophecy project needs more definition as to what it is and a transition to the next sentence.[17] COVID-19 studies attempt to identify clusters based on travel histories, individual movements, and tweets. (e.g., satellite images of industries and companies, environmental impact).

Alternative data means an attempt to know anything around the business based on what "others" think of the business. For example, if the public health wants to know the path of the pandemic, it will source exotic data such as parking lots and cell phone usage.

Alternative data provides collective intelligence resulting in rewards and risks. Risk arises from the source of data, which remains unverified.

Risks in alternative data usage are worth in time-sensitive business processes. For example, in financial trading, even a narrow timing advantage provides a trading edge. Alternative data provides that edge. Alternative data provides potential information advantages in investment management decisions.

Alternative data is usually incomplete and not verifiable. They are unstructured and difficult to integrate with existing organizational, transactional data. Privacy and security of this data are also very difficult to establish.

Regulation also plays a role in sourcing and use of alternative data. For example, with GDPR regulation, it may be necessary for an organization to explain and make visible the alternative data they have collected from community forums and which they are using to identify trends in a product or service. In other regions, where GDPR does not apply, the need to disclose the alternative data source and usage is not required.

Data security and storage

Security and privacy issues for large volumes of data increase with their increasing use in analytics. This challenge of security increases further as the velocity of the data being generated is also very high. Cloud-based storages reduce the pressure to plan for data storage infrastructure. Cloud storages do have the challenge of ownership and security of data. Additionally, in many cases due to the velocity of data, real-time backups may not be possible. Data redundancy and backup of data are other important security and storage factors.

DATA ANALYTICS IN BUSINESS PROCESS OPTIMIZATION

Analytics themselves are not new. Starting with an abacus to the use of a sophisticated Excel with macros, analytics provide valuable insights for users. Analytics are a pointer to (a) what action is to be taken and (b) the results of that action.

Data analytics

Analytics provide a data-centric approach to business decision-making. Analytics have multiple purpose and importance. Analytics explore the past by describing a happening based on static historical data. Analytics also look forward by presenting what is likely to happen in the future – a prediction. Predictive analytics need to incorporate dynamic data inputs and use them to create multiple "what-if scenarios."

Data analytics is at the heart of business optimization. The starting point of a strategic approach to using analytics in BO is the business problem or desired outcome by using those analytics. Examining the factors contributing to value creation in business is more important than the details of analytics and technologies. Once the business outcomes are understood, data analytics start examining the diverse data sources and types available to the organization for analytical purposes.

Data analytics generate insights that are translated into actionable knowledge. Data analytics includes clustering, segregation, segementation, and analysis of the customer data in order to understand customer behavior, potential sentiments, and referrals. This analysis results in actions. Enabling those actions to happen requires changes to business processes and people training. For data analytics to succeed, a clear roadmap for its application is essential.[18]

Business process optimization

Optimizing business processes is making efficient and effective decisions. Organizational functions generate data and use data analytics. Project management, enterprise architecture, process modeling, solutions development, and quality assurance are a number of organization-level activities required for successful optimization. These activities complement data science. Categorizing data, finding the correlation, undertaking analytics, and presenting the insights (in an easy-to-use way for the end-user) result in improved decisions. This improvement in decisions is measured in terms of speed, accuracy, and points where decisions are made.

The following activities are undertaken for business optimization:

- Specify, at a high or strategic level, the desired business outcomes from the optimization effort. This is a leadership activity that sets the business vision from AI in close collaboration with the senior leadership team.
- Identify current capabilities and technology maturity of the organization of an enterprise. This requires a survey of data inventory.
- Identify current business processes and how these processes will change (reengineered) with AI.
- Start establishing the context in which data will be used.
- Also, identify gaps in the current business processes and the ones that can be data-driven. This is achieved through process modeling.
- Understand the risks of AI on business processes, including the change to organizational decision-making and structure.
- Decentralize decision-making in the business processes requiring mindset change.
- Document operational and performance parameters used in measuring user/customer satisfaction.
- Create process optimization models with the end-user in mind.
- Categorize current structured data both within and outside an organization and how it is impacting current decision-making. This information is helpful in bringing together the structured and unstructured data to provide a holistic, 360 degree view to the customer/user.
- Approach to integrate semi- and unstructured data with existing structured data in order to enhance analytical insights in order to reduce the risks associated with such integrations.
- Develop a strategic approach to the use of alternative data. Include alternative and data analytics in decisions.
- Explore the correlations of business functions in an organization (e.g., between revenues and marketing, products and customer

satisfaction, skills and project success). Data analytics engenders such correlations by relating information from otherwise disparately spaced data silo.

- Work with the limitations of existing data-related skill levels across the organization and help formulate an upskilling approach. This is aimed at enhancing the capabilities of the current staff by either providing training and experience, or supporting them with external consulting resources in the initial stages of adoption.

Establishing the data context

Developing the context and using context awareness are vital in Big Data analytics. For example, each IoT (e.g., wristwatches, smoke detectors, shoes, and home appliances) are a data point with many, additional data points embedded within them. These devices can send, receive, and process data in collaboration with other devices and the back-end Cloud in real time. Embedding a sensor, sending signals, and receiving data points over the Internet is a start. This start is followed by context awareness of data from multiple sensors coming from varied sources using advanced algorithms, in real time, to develop a 360-degree, holistic view of the data point for enhanced and Agile decision-making.

Typically, although not always, the context can be ascertained by who, why, what, when, where, and how. A simple IoT device, with limited functionality, may only need to answer one or two W's, while a more complex device generating complex sensor data may need to answer all questions. Table 2.2 shows an example of contextual reference for a "patient" data point when used in data analytics.

Tools and techniques for BO

Big Data technologies (build on Hadoop[19]) ingest and store data in different formats including unstructured data such as customer feedback (via emails, blogs, and forums).

Big Data based on Hadoop (HDFS) is primarily static, batch-oriented, and analyzable. The architecture of Hadoop had not incorporated the velocity of data and its real-time processing. Therefore, additional technologies are required to handle the high-velocity data. This is so because irrespective of the incoming format and velocity, eventually all data points need a semblance of structure in order to be analyzable.

Additional data manipulation tools and techniques are required to interface incoming transactional as well as unstructured data with the large, static enterprise data warehouses. Ease of configurability and ease of use of these tools and techniques play an important part in value generation from analytics.

Table 2.2 Parameters of a data point (patient) in ascertaining business outcome

Context (reference point)	Description of the reference point	Example of the contextual reference point for "patient"
Who	Stakeholder	Patient in a hospital around whom the entire context is based.
Why	Goal	The patient stakeholder has the goal of getting well and that provides input into the context of storing and analyzing this data point.
What	Tools and technology	Hospital bed, ventilator, monitors, and associated tools provide the technologies for the patient data point. These tools and technologies provide the input into the speed with which the "Why" or the goal is achieved and the costs.
When	Timing	I AM provides input in the context because it changes the way in which the patient is transported, and the speed with which the treatment is made available. A macro-level context is whether the patient is coming when the hospitals are full due to a pandemic.
Where	Location	Specific location of the hospital, the bed, and whether the patient is taken straight into the ICU because of an emergency.
How	Process	Admitting the patient and providing the treatment to enable the patient to achieve the goals.

Analytical tools and techniques help in handling the challenges – particularly when handling unstructured data. For example, analytical techniques map and index a suite of unstructured data to identify a pattern in the data. Statistical analytical techniques (e.g., linear regression) are complemented by software and programming tools (e.g., "R"[20] or Python[21]) and solutions packages (e.g., SAS[22]). NoSQL is the domain of aggregate-oriented databases[23] that provides increasingly sophisticated mechanisms to store and analyze unstructured data. Tools for data security analytics are discussed in Chapter 9.

Data analytics design for BO

Designing data analytics is a balancing act – provisioning enough to let the users configure the analytical solutions they want but at the same time ensuring enough integration and control in the background to ensure security, accuracy, and ease of use. The Agile concept of iteration and increment is invaluable in developing such balanced, configurable self-serve analytical solutions. While it personalizes the experience for the user it also frees up valuable organizational resources and enables lean business processes.

BO is reliant on the richness of analytics, the numerous "what-if" scenarios, and their timeliness. The volume, variety, and velocity with structured, unstructured, and semistructured data types are studied together in data analytics. The complexity of such data is acknowledged as a challenge and also as a promise for the insights it can produce. For example, data analytics goes beyond sampling and discrete categorizing and into the realm for analyzing full datasets. This is a challenge as well as an opportunity to produce very precise results.

Granularity of analytics in BO

Another important concept that helps business agility is the levels of granularity in undertaking analytics. Granularity of data, granularity of analytics, its context, and the processes embedding the data are crucial in establishing data strategies for Agile business.

Big Data analytics takes analytics to a very fine degree of granularity. This finer granularity is enabled by algorithms turned into code. The execution of this code on the high-volume and velocity data is enabled through the technologies of Hadoop and NoSQL. Finer granularity of analytics is the capability that differentiates Big Data analytics from the traditional analysis of data.[24] The availability and accumulation of data combined with the availability of computing power enables drilling down through that data with pinpoint accuracy. Ascertaining the optimum level of this granularity of data analytics is a strategic business decision discussed later in this chapter.

Velocity coupled with volume requires strategies for handling data that also start with the desired business outcomes. For example, the more finely granular are the analytics, the greater is their confidence level although more resources are required from the organization. This, in turn, can increase the cost of analytics. Coarse (or lesser) granularity means less precision in the results. Granularity and resources are thus a continuously balancing, "agile" act. Desired outcomes enable the establishment of the right levels of granularity.

User experience analysis and BO

Another important aspect of data analytics is understanding customer (user) sentiments through "user experience analysis." Data generated by social-mobile interactions before the user is in direct contact with the business is explored as part of user experience analysis. This exploration enables a business to come up with business strategies that aim to understand the customer before and after the contact period of that customer with the business. This data can be analyzed in order to understand the customer expectations and the user behavior and set the business response accordingly. User experience analysis depends heavily on social media and mobile technologies.

Analytics provide for growth and innovation in services and enable process optimization. A strategic approach is a must for data analytics to be part of the organization culture. Precision and speed in analytics and interactive data exploration need to be supported by upskilling the users and providers of the analytics.

Self-serve analytics in BO

Self-serve analytics (SSA) is a business strategy focused on letting users (e.g., staff and customers) decide what they want from the DatAnalytics (data-cum-analytics), how they want it, and then help them achieve those insights. This not only provides the user with what she wants, when she wants it, and how it is delivered; but it also reduces the onus of analytics on the business.

Given the velocity of data, the rapidly changing context of the user (e.g., the urgency of the results, the format in which they are desired, and the device on which they will be presented), and the ever-increasing availability of data warehouses from "third parties," it is in the best interest of an organization to provide patterns or prefabricated analytical tools which enable users to self-serve themselves. This is called "Customer Intelligence."[25] Self-serve analytical capabilities require back-end data integration, dynamic business process modeling, and customizable visuals. Tools are used to bring data together from many different locations – internal and external – to instantaneously answer self-service queries. These tools enable collaborations among multiple systems, their interfaces, and open data sources from external organizations (e.g. third-party sites providing data or that being sourced from data providers).

Gartner[26] describes the ease of use as the topmost priority for self-service analytics. This is because the user is unlikely to have technical or analytical skills. The user could be a banking or financial analyst and not necessarily a statistician. Visual configuration of services needs to be presented to the user. These users should be able to configure analytics to solve their immediate needs.

DATA CLUSTERS AND SEGMENTATION

Clustering groups similar data together into clusters. This clustering is done with the aim of segmenting them. Segmentation places data into groups based on similar characteristics. These terms denote different approaches to analyzing data. ML algorithms help identify clusters by identifying relationships of different types of data. Clustering deals with finding the relationships between data that are then placed in segments. Clustering the data helps discover new segments.

Horizontal and vertical clustering

Clustering identifies groups of similar entities in a multivariate dataset. These are business data such as customers, products, sales, marketing, and regions. Clustering is undertaken through partitioning of data in a horizontal and/or vertical manner. Good clusters are based on internal cohesion but externally loose coupling. Minimal differences within each cluster but maximum differences across clusters is considered as a good analytics design.

Segmentation

Data segmentation forms the basis for process optimization. Segmentation is the separation of data into groups of similar elements. This is an important first step in analytics. Age vs gender or customers vs prospects are examples of segmentation. Segmentation allows businesses to analyze customers based on those similarities. Segmenting enables the creation of a target for services or products. For example, a data analyst helps new businesses improve their decision-making capabilities by segmenting other similar businesses. Data from these businesses is further divided for further analysis.

The process of segmentation continues in an iterative manner with increasingly finer segments. Segmentation finds relationships within variables in order to predict customer behavior.

Clusters and segments in practice

Segments are based on relationships between datasets. However, there is a limit to establishing these relationships. Clustering algorithms identify similarities in data. The closer the data elements are to each other, the better the segment. Distance between data items should be less but more between segments. Clustering can primarily lead to unsupervised learning in analytics. Clustering is an example in unsupervised learning, wherein the model has no target variable. Clustering groups the data based on multiple variables without being explicitly labeled.

Instead of grouping customers, clustering identifies what customers are likely to do.

ML utilizes clustering and classifications based on data features in practice. Data analytics drives and optimizes functions such as marketing, sales, support, supply chains, operations, and HR. Users of products and services benefit with personalization of services. Analytics benefit with personalization of services. Segment, cluster, and aggregate analyze data. Segmentation and clustering help prepare data for cluster analysis for business optimization. Here are some examples in practice:

- Market – Clustering potential customers with similar interest in a product based on their similarities.

- Sales – Clustering can tell what kind of people buy a specific product.
- Insurance – Clustering techniques are used to identify fraudulent insurance claims.
- Education – Identifying groups of universities based on their tuition, geographic location, quality of education, and type of degree programs.
- Credit – Grouping customers based on their credit history.

DATA-DRIVEN DECISIONS

Nature and types of decisions

Figure 2.6 shows a simple matrix of AI and NI. Automation, prediction, experience, and intuition are the four quadrants that fit in the known-unknown aspects of this matrix. These quadrants form the basis for the current discussion on enhancing the AI-based tools to support decision-making by leaders. The prediction and the experience quadrants of Figure 2.6 are of greater interest in the context of this discussion as compared to the automation and the intuition quadrants. AI-based predictions and leadership experience is a judicious combination for BO.

Automation

Automation happens in the ML space wherein machines take over routine processes in order to reduce time and effort for humans. Any process that is primarily simple, straightforward, and deterministic in nature requiring

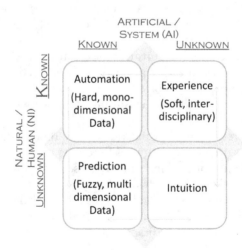

Figure 2.6 Known-unknown disruptions matrix for AI versus NI.

minimal adjustments during execution is ideal for automation. Automation works best in handling a known or definable disruption that has occurred in the past, and the decisions and actions taken at the time of the occurrence are recorded within the AI system. For example, known network failures in the technology space benefit by automation. Monitoring networks and network communications is a likely beneficiary of automation because the processes around network management are usually well defined.[27] As argued by Gonsalves and Unhelkar (2020), the primary purpose of automation is to improve the performance of a system.[28] This performance is usually measured in terms of time and costs – both of which are improved through automation. While this aforementioned improvement in performance may be considered as "value," it is not the same value as perceived by human customers (See Chapter 10 on NI). Disruptions to businesses result not only in the loss of time and money, but they also result in the loss of value as perceived by the customer/user. Situations, such as COVID-19, which are not amenable to prediction with simple and deterministic equations, are unlikely to benefit by automated systems. Disruptions are complex; they comprise technology, processes, and user disruptions that require complex stakeholder management and recovery strategies. Disruptions cannot be handled entirely by automation, which is limited to the measurable aspects of a business system – primarily time and costs.

Prediction

Prediction occurs when the disruptions can be anticipated based on trends in data.[29] AI-based predictions can be categorized as "known" to the system but "unknown" to the human. This is so because the tools and technologies of AI are well equipped to make sense of the vast amount of fast-moving data. Leadership, in this category, needs to rely on the AI-based system to sound the alarm when there is a possibility of disruption. Big Data analytics have shown to predict network failures and security breaches based on the analysis of large amounts of historical data.[30] Since predictions improve with data, the larger the dataset, the better the predictions. Economic disruptions, for example, are predicted with reasonable confidence based on data trends. ML code embedded in AI models is, however, agnostic to the specific situation, or context, of the user.

Experience

This represents the human-based NI that comes into play in decision-making. AI engines work only on available data which, in some situations, may not be readily available. In other situations, the tools may not be able to identify corresponding trends within a reasonable time. When the situation is unknown to the system, leaders provide the needed input based on their experience.

NI can undertake anticipation, decisions, and actions based on leadership experience. For example, where systems fail to predict network failures or security breaches, technology leaders anticipate these events based on their experience. Resilience and recovery strategies are intertwined in the very thought processes of experienced leaders. Thus, while due consideration is given to the predictions provided by the automated ML modules, these predictions are superimposed with NI.[31] Despite the cutting-edge deep learning algorithms, human experience is not overridden, and decision-making is not entirely handed over to AI. Furthermore, experience-based anticipation and actions include consequences of the decisions in relation to quality and ethical ramifications. These aforementioned consequences and the human expertise (NI) are important considerations in the "known to humans, unknown to system" box in Figure 2.6. The learning algorithm combines its historical analytics with the freshly incoming data in order to improve its predictability of disruptions. The learning algorithm undergoes continuous improvements through ongoing iterations and increments in an agile manner.

Intuition

This concept encompasses the space where disruptions neither have a past – historical data to create a trend – nor do they appear in the experience horizon of the leaders. In other words, these disruptions are unknown to AI and unknown to humans – as shown in the right-bottom quadrant in Figure 2.6. The current COVID-19 pandemic falls under this category of unknown-unknown. The only possibility of anticipating such disruptions is to explore the instinctive feelings within the leaders, together with imaginary scenarios created in the AI engine.

DATA ANALYTICS FOR BUSINESS AGILITY

An optimized business is an agile business. Optimization is achieved by continuously enhancing business processes with the help of data analytics. Figure 2.7 summarizes how an optimizing business leverages data analytics for business agility.

- Customer analytics enables the business to understand the changing preferences of the customer. Personalized needs of the customer are also better understood through this analysis.
- Product analytics provides insights into the costs, risks, and relevance of the product to the users. Processes related to the development of the product are optimized using this analysis.
- Marketing analytics enables the business to understand the market gap, its size and location, and how to approach the promotion of its products or services.

Figure 2.7 Leveraging analytics for business agility.

- Sentiment analytics is an understanding of the customer sentiments after the product or services has been consumed. It is a useful tool to figure out if the customer is likely to support and promote the business or work against it.
- Security and privacy analytics is the understanding of the confidence level of customers and other users in the products and processes of the business.
- Cross-selling analytics establishes the correlations between the various products and services of the business as well as its partners and competitors.
- Maintenance analytics is invaluable in a manufacturing setting in particular as it provides descriptive as well as predictive insights into the state of the equipment.

The analytics alluded to above impact the following business functions.

- Customer segmentation and attrition is optimized due to a better understanding of the customers, their needs and priorities, and changes to the customer groups as they move their priorities.
- Product acceptance on user community is an interesting process that deals with the generation and use of alternative data. This is the acceptance (or

lack thereof) of the company's product or service on the user community platform.

- Targeted marketing campaigns on multichannels can be brought about with the help of the analytics mentioned above.
- Social media discussions and blog posts can be influenced by active participation of the organization in those discussions and alleviating the issues faced by the customers.
- Recording search and transactional data in order to further update the databases and fine tune the analytics.
- Strategizing for cross/up-selling the products and services is enhanced and focused as a result of the use of analytical insights.
- Pre-empting events such as the launch of products or, in the worst case, redressing large-scale grievances can be organized based on analytics.
- Optimizing the business processes of the organization by having a better understanding of the slack within them and how they are providing value to the customer

CONSOLIDATION WORKSHOP

1. Why are the four "sets" of *Think Data* important?
2. How is AI positioned within the five layers of the Data to Decision pyramid?
3. Data is a set of observations which may be influenced by the subjectivity and perceptions of the observer. What precautions need to be taken to ensure objectivity and bias-free data?
4. What are the similarities in AI utilization from Data to Big Data? What are the differences?
5. What are the challenges of handling the categories of Big Data and the 4+1 "V" of Big Data in the context of business?
6. What is Alternative data? Discuss the characteristic of this type of data and its importance in BO?
7. What are the challenges in making sense of Big Data in business decision-making? How can these challenges be addressed?
8. How can the types of Big Data Analytics provide opportunities for business?
9. What is the difference between Segmentation and Clustering of Big Data and the impact of each within ML algorithms?
10. What are the advantages and nuances of data-driven decision-making?
11. Describe the challenges and risks in data-driven decision-making.
12. What is the "known-unknown" matrix for AI and NI? Discuss with examples. Explain the role of experience and intuition in the "known-unknown" matrix for decision-making. How can they be interfaced with AI?

13. Can AI/ML algorithms figure out the context in which they are applied?
14. Describe the role of learn-build-adjust cycle in the feedback loop of refining decisions in the data to decisions pyramid.

NOTES

1. https://www.statista.com/statistics/871513/worldwide-data-created/ accessed 17 Oct 2020.
2. Hazra, T., and B. Unhelkar, 2020, *Enterprise Architecture for Digital Business: Integrated Transformation Strategies*, (CRC Press, Taylor & Francis Group) (in Production).
3. Tiwary, A., and B. Unhelkar, 2018, *Outcome Driven Business Architecture*, (CRC Press, Taylor & Francis Group /an Auerbach Book, Boca Raton, FL) Co-Authored.
4. Satellite imagery of car parks in Wuhan province as reported: https://www.bbc.com/news/world-asia-china-53005768 and https://www.cnn.com/2020/06/08/health/satellite-pics-coronavirus-spread/index.html.
5. Unhelkar, B., 2013, *The Art of Agile Practice: A Composite Approach for Projects and Organizations*, (CRC Press, Taylor & Francis Group /an Auerbach Book, Boca Raton, FL) Authored ISBN 9781439851180, Foreword Steve Blais, USA.
6. Sadalge, P., and M. Fowler, *NoSQL Distilled*, Addison-Wesley, p. 3013.
7. See note 5
8. Unhelkar, B., and A. Sharma, 2017, March "Innovating with IoT, Big Data, and the cloud", *Cutter Business Technology Journal*, Vol. 30, No 3, pp. 28–33.
9. Helmy, M., A. Arntzen Bechina, and Arvid Siqveland. 2018. "Using machine learning for identifying ping failure in large network topology", In *Economics of Grids, Clouds, Systems, and Services*, edited by Emanuele Carlini, Massimo Coppola, Daniele D'Agostino, Jörn Altmann, and José Ángel Bañares. *5th International Conference*, GECON 2018, Pisa, Italy, Sep 18–20, 2018: Springer International Publishing.
10. Unhelkar, B., and A. Arntzen, 2020, "A framework for intelligent collabora- tive enterprise systems: Concepts, opportunities and challenges", *Special issue of the Scandinavian Journal of Information Systems*, Ed. Eli Hustad, Summer.
11. Unhelkar, B., 2017, May 1 "Overcoming the Big Data strategy lacuna", *Cutter Executive Update*, Data Analytics & Digital Technologies Practice, Vol. 11, No. 11.
12. Unhelkar, B., 2018, *Big Data Strategies for Agile Business*, (CRC Press, Taylor & Francis Group/an Auerbach Book, Boca Raton, FL). Authored ISBN: 978-1-498-72438-8 (Hardback).
13. Günther, Wendy Arianne, Mohammad H. Rezazade Mehrizi, Marleen Huysman, and Frans Feldberg, 2017. "Debating Big Data: A literature review on realizing value from Big Data", *The Journal of Strategic Information Systems*, 26, pp. 191–209.
14. See note 12.

15. IoT is a paradigm that is based on the device suite called the ADC – from Pradhan, A., and B. Unhelkar, 2020, "The role of IoT in smart cities: Challenges of air quality mass sensor technology for sustainable solutions", *Chapter 13 in Security and Privacy Issues in IoT Devices and Sensor Networks*, edited by Sharma, S., Bhushan, B., and Debnath, N. C., Elsevier, Book Series: Advances in ubiquitous sensing applications for healthcare, ref. B978-0-12-821255-4.00013-4.

16. http://www.informationweek.com/mobile/mobile-devices/gartner-21-billion-iot-devices-to-invade-by-2020/d/d-id/1323081 accessed on 21 Nov 2016.

17. Rickards, J., 2017, *Death of Money: The Coming Collapse of the International Monetary System*, (Penguin Publishing Group).

18. Agarwal, Anurag, Ramkrishna Govindu, Sunita Lodwig and Fawn Ngo, 2016, April "Solving the jigsaw puzzle: An analytics framework for context awareness in IoT", *Cutter IT Journal*, Vol. 29, No. 4, pp. 6–11, Ed. B. Unhelkar and S. Murugesan, Special Issues on IoT Data Management and Analytics: Realizing Value from Connected Devices.

19. Hadoop and distributed database architecture - https://hadoop.apache.org/ accessed 29 Oct 2020.

20. www.r-project.org/ accessed 29 Oct 2020.

21. https://www.python.org/ accessed 29 Oct 2020.

22. www.sas.com accessed 29 Oct 2020.

23. See note 6.

24. Based on Agarwal, A., "Predictive analytics and text mining", *Presentation at The Suncoast Technology Forum*, Sarasota, FL, 16 Feb 2016.

25. A SAS white paper "Best practices for delivering actionable customer intelligence a TDWI checklist report", http://www.sas.com/en_us/whitepapers/tdwi-delivering-actionable-customer-intelligence-107984.html accessed 27 Nov 2016.

26. Gartner. "Survey analysis: Customers rate their advanced analytics platforms (G00270213)", 2014, Oct 28.

27. Boutaba, R. and J. Xiao, 2002, "Network management: State of the art", *Communication Systems. IFIP WCC TC6 2002. IFIP — The International Federation for Information Processing*, edited by Chapin L., Vol. 92. Springer, Boston, MA.

28. Gonsalves, T. and B. Unhelkar, 2020, May. "Superimposing Natural Intelligence (NI) on Artificial Intelligence (AI) for optimized value in business decisions", in *Intelligent Automation: The Gateway to Optimizing Business Performance*, edited by Murugesan, S., Cutter Business Technology Journal, Vol. 33, No. 5, pp. 26–32.

29. Agrawal, A., J. Gans, and A. Goldfarb, 2018, *Prediction Machines: The Simple Economics of Artificial Intelligence*, (Harvard Business Review Press).

30. Arora, D., K. F. Li and A. Loffler, 2016. "Big Data analytics for classification of network enabled devices", *2016 30th International Conference on Advanced Information Networking and Applications Workshops (WAINA)*, Crans-Montana, pp. 708–713, Doi: 10.1109/WAINA.2016.131.

31. See note 28.

Chapter 3

Digital leadership
Strategies for AI adoption

STRATEGIZING FOR BUSINESS OPTIMIZATION

Digital business is based on data as a strategic asset. Strategic planning for business optimization (BO) incorporates data, analytics, and application. Providing customer value is the purpose of utilizing data. Artificial intelligence (AI) impacts organizational culture, business process, systems, and data. Eventually, AI adoption impacts the customer. Digital business strategy for BO depends on the context. A market, product, process, and government policy are examples of the context. Furthermore, automation, optimization, and humanization play an important role in digital business strategies. Data usage also leads to new and opportune business models. Data-driven analytics in business processes enable data-driven decisions, innovation in products, and efficiency in business functions. BO expands the use of data beyond IT and into the rest of the organization. BO improves business performance by enhanced decision-making. Decisions around operational optimization are based on data-driven analytics. Such digital strategies lead to the discovery of new means of delivering customer value.

Adopting AI-based decision-making involves both technical and business risks. Leaders understand and incorporate those risks within digital business strategies. Costs, returns, risks, security, privacy, and safety associated with AI-based BO are outlined within the business strategy. Development and deployment of AI solutions, reengineering of processes, and change management form part of business strategy. Digital business strategy also ensures the security of business data from a regulatory and compliance viewpoint.

Business strategies for optimization treat AI adoption as a holistic, organizational challenge rather than a project-based initiative. AI for business optimization is also not considered a technology-based initiative. Instead, digital strategies are developed to take a long-term and holistic view of AI adoption in the organization. Apart from collaborating in the digital world, businesses are continuously merging and acquiring each other. These mergers and acquisitions lead to the challenge of data and technology silos. Planning the optimization of a business includes planning for integration and use of data across multiple organizational boundaries.

Envisioning digital business strategy for AI

Digital business strategy encompasses visions, goals, capabilities, and projects. The strategies are aimed at developing business capabilities. While projects play a role in developing the capabilities, projects themselves are not the mainstay of the BO effort. Digital implementations are accomplished through projects but their purpose is to develop capabilities for the business.

A digital business strategy for AI incorporates the following elements:

- Vision and aspirations of the business and understanding how AI influences them. These business visions provide the starting point for the BO effort. Providing customer value is the basis of this vision for most digital organizations.
- Defining goals that enable business outcomes based on AI capabilities. The vision provides the guiding principle behind the development of goals.
- Identifying and defining the capabilities of the business – these include the current and future (required) capabilities for data-driven decision-making in the business.
- Planning the implementation of capabilities in business processes through the projects.
- Developing a risk management plan that considered the risks associated with business optimization, and the risks in the business that can be handled with the help of AI-based technologies.
- Outlining the relevant metrics and corresponding measurements that will help in justifying the BO initiative as well as in reporting on its progress.
- Planning for cybersecurity and privacy considerations to be incorporated in BO.
- Due consideration to the people's concerns of job losses due to automation and optimization.
- Optimization has the potential of customer perception of less personalized service which needs to be considered strategically. Optimization with AI implies change which is a risk. Strategies balance new process implementation with current operations.
- Seeking regulatory approvals in terms of use of data and processes.

The vision and goals of the organization provide the basis for desired outcomes in BO. Prioritized outcomes influence the digital business strategy and refine vision and goals. This iterative approach to vision and prioritization continues to refine the strategy. Strategies stabilize after a couple of iterations. Strategies reduce changes in direction. Processes are defined, measured, and implemented as part of BO. The implementation of digital strategies is then assigned to business units. The business strategy generates actions across the organization.

Digital strategies are holistic

A digital business strategy provides the necessary guidance and control in optimizing business processes. Such digital business strategy includes multiple disciplines and functions of the business. A good data-driven strategy works to integrate data with processes and people with due consideration to governance, security, storage, and usage. Digital strategies make provision for the inevitable collaboration and merger of organization that require the unification of data and modeling of collaborative business processes.

Data unification across multiple processes and businesses enables the creation of a singular customer view, single point of contact of customer with the organization, enhanced quality and governance, and greater collaboration between technical and business stakeholders. Technical and business silos are melted by the holistic, long-term view of the strategies, resulting in a unified view of the customer. Integrated data also leads to efficient analytics and accurate insights as it eliminates the need to process the same data multiple times and focuses on a singular view of complete and accurate data. The integration of data is thus invaluable in data analytics and BO.

The vision and mission of the organization provide the necessary direction and impetus in developing BO strategies. The vision, goals, objectives, and corresponding requirements specific to AI are brought together with BO strategy and planning. AI, data, and infrastructure of the organization play a supportive role in the achievement of the goals. Data-based AI technologies present myriad challenges during optimization, including and especially security and privacy of data and users. Digital strategies prioritize security requirements in terms of risks, issues, and concerns across the cloud, Big Data, and IoT implementations. Evaluation and assessment of the existing business strategy is a helpful input in the development of the new digital strategy. Digital business strategy considers an optimal approach that may work across the enterprise – with a caveat that it may or will change during the course of transformation – together with a plan to handle the risks.

Business leaders who develop the digital strategies pay close attention to the culture and values of an organization. The cultural aspect of an organization is its mindset within *think data* discussed in the previous chapter. The mindset is the most challenging aspect of change within BO. Digital strategies give significant importance to the "soft" issues of culture and values. Leaders take a strategic rather than tactical view of optimization beyond automation. Tactical use of data is primarily in automation. Strategic use of data includes removal of redundant activities in a process, creation of new collaborative processes, and ensuring business agility.

Customer value is the goal

Providing customer value is the key goal of BO. A consistent and succinct roadmap for BO focuses on the customer experience. Enhanced customer experience includes engaging the customers, offering a wide range of business services, and providing better quality and timely responses.

BO also changes the organizational structure and behavior as processes are reengineered and decision-making is decentralized. These changes impact the competitive position of the organization. Metrics are essential to monitor the implementation of the digital strategy. For example, satisfaction, loyalty, and cost metrics provide implementation control and direction. The business architecture keeps customer, infrastructure, and resources in continuous alignment. Strategies are continuously aligning the capabilities to achieve business outcomes.[1]

Addressing the business goal or problem

A business problem is a good starting point for developing BO strategies. Cumbersome or unsecured business processes may result in lost customers. Consider, for example, a business problem of rise in customer complaints for billing errors. These operational problems can have an immediate, short-term fix. However, BO examines the entire business and re-engineers it not only for the immediate problem but a suite of known and unknown problems. An upgrade of the software system with embedded analytics or establishing a call center equipped with AI can be considered a strategic-technical solution. Prototyping the solution and enabling key business stakeholders to provide input make it relevant to the business.

Business problems are investigated and analyzed strategically in BO. Agility plays a positive role in enabling this investigation in a trustful, honest, and collaborative manner.[2] Investigating business problems provides strategic options in using AI. A composite Agile approach that balances formal planning with agility provides an excellent opportunity for developing and deploying AI in business.

Business agility in decision-making

Digital business strategies aim to use data-driven decision-making across every business function. Developing strategies to optimize operational areas of business is not a one-off deliverable. Instead, developing BO strategies is also an iterative and incremental approach that benefits with the use of agile principles and practices.

Agile as a business characteristic carries immense value.[3] The agility goals of business are included in the strategic planning effort. The balance between customer focus and operational continuity is achieved only with business agility. In turn, this strategy depends on an Agile approach to knowledge management and project management.

The moment agility in processes is introduced, it brings in elements of iteration and increment that aim to handle the uncertainty and change in the process.[4] Even though ML algorithms are now able to mimic human behavior to a large extent, the behavior is itself a function of widely differing values for the same input. An optimized business makes faster decisions that are more accurate than before, and that are made over a larger number of touch-points. Therefore, the more the iterations (agility), the better are the chances of anticipating and ameliorating disruptions.

Following are agile-specific considerations in developing a strategy for BO:

- Identifying the opportunities for products and services innovation in terms of their processes, creating their prototype, and experimenting with the prototype in order to provide input in the strategies
- Identifying, categorizing, and prioritizing the risks in optimizing business processes with AI and incorporating those risks in the strategy
- Using Agile iterations to understanding and scope the overall requirements of the BO initiative and directing investments in AI accordingly
- Modeling the parameters around providing customer value and using those parameters as input for the development of strategy
- Digitizing the external (marketing and sales) functions of an organization and planning to incorporate them in social media in an iterative and incremental manner
- Aiming to optimize the internal (operational) business processes with analytics and AI in an iterative and incremental manner
- Ensuring governance, risk, and compliance of all business functions digital reporting of compliance requirements using the principles of Composite Agile Method and Strategy (CAMS)[5]
- Applying the principles and practices of cybersecurity to the digitized business processes in a holistic manner
- Keeping the metrics and measurements to the bare minimum at the start of the initiative and slowly enhancing and enriching the measures with the initial iterations of implementing AI

STRATEGIC PLANNING FOR BO

Strategic planning for digital business is the creation of a plan for the digital business to achieve its vision and goal. Such plan also focuses on the alignment of business functions with all service and support functions, especially IT. The strategic plans, goals, objectives, and expectations of the digital business are periodically synchronized with capabilities and project initiatives of the organization. Strategic planning identifies goals, builds a roadmap, and develops a blueprint for implementing change. Modifications and upgrades to IT capabilities result from strategic planning which aids in the transition to a digital business.

Strategic planning is also associated with qualitative and quantitative measures. The metrics and measurements enable buy-in from the business for the digital transformation. The interactive, innovative, and collaborative coalition of IT and business has a significant implication towards delivering the right business results.

Strategic planning for BO considers the following impact of the changes in the decision-making process:

- Current Decision-Making Process – this is important to understand as it will have gaps in delivering value to the customer. Any changes to the decision-making process are based on the current process and the risks associated with changing it based on data-driven decisions.
- Impact of Data on Decision-Making – including the sources, storage, and manipulation of data. Data sources from collaborative business partners, third-party and government sources, and the in-house generated data stored on the Cloud impact decision-making and are included in the strategic planning process.
- Impact of Analytics on Decision-Making – especially as this impact will change (improve) the accuracy and time taking for making decisions.
- Impact of People and Biases on Decision-Making – strategies need to factor in the changes that digital business brings about on people and the way they make decisions. Changes associated with human bias and sensitivity require careful strategic planning as these, perhaps, are the most important and difficult parts to change in a DT.
- Business Process Reengineering – when undertaken while the organization is still in operation requires substantial and holistic strategic planning. This is so because the impact of digitizing a process on another process and functions of the organization can be substantial.
- Collaborative Impact – especially because DT will render the business much more collaborative on the electronic platforms than ever before. Strategic planning needs to include due considerations to the services offered and consumed from collaborative business partners, policies governing those services, and the impact on the value being delivered to the customer.
- Cybersecurity – is a vital strategic consideration in DT as changes on the electronic platform open up the organization to the possibility of cybercrime and related risks. Data, analytics, and digital business assets need additional protection and those protections need to be factored in, in strategic planning.

"Think data" in strategies

Strategies ensure AI plays an important yet supportive role in business decision-making. AI is an enabler and an agent for change. Data is a business asset that provides value. *Think data* is a holistic approach to discovery of that value. Extracting knowledge from data requires a strategic approach.

Business optimization is the application of *think data* in effective utilization of limited resources and capabilities. Data thinking is the extraction of useful knowledge and business value from the data. A strategic approach brings close collaboration between business stakeholders and data scientists.

Business architectures[6] help business organizations achieve effective utilization of resources, as well as generating customer value. Data and architectures based on 4+1 V (Volume, Velocity, Variety, Veracity+Value) of Big Data help align AI solutions with BA based on Big Data strategies for Agile business.[7] Big Data and AI are strategically important in optimizing business processes.

Machine learning (ML) systems learn from previous decisions and provide insights into those decisions for future decisions. Big Data provides vast repositories of data (decisions and the data on which those decisions are based) in order to improve the speed, accuracy, and reach.

Collaborations between businesses need to be strategically thought through. Organizations are brought together by digital platforms. Viewing the entire organization as a whole is essential for developing business strategies.

Effective use of data creates a fine granular decision-making engine for multiple decision points. A well-constructed and well-maintained strategy leads to a lean and agile, digital business. Lean processes enable a business to update its strategies to correspond to the changing business environment.

Strategic AI considerations

Strategies describe the actions needed to achieve business goals. AI is a part of the enablers to provide customer value. Risks associated with the implementation of BO are described and managed through the strategic plan. Projects are aligned to the desired business outcomes.[8]

Strategies based on the business outcomes make the businesses customer-centric. A business strategy to improve customer interaction by optimizing supply chains used AI to optimize the entire process. An online business fulfills fast-moving consumer goods through AI. Products are offered based on optimized supply chains. Strategies also protect innovations from being mimicked by the competition. Sound business strategies incorporate the AI technologies in business processes, pay attention to customer needs, and provide continually balancing implementation of AI.

Composite Agile Method and Strategy (CAMS)[9] professes a comprehensive and strategic approach to developing organization-wide agile capabilities that include people–process–technology–money aspects of the entire business. This strategic approach is also applicable to AI adoption. While AI adoption across an entire organization is more complex than implementing a single AI solution, the rewards are correspondingly greater. Organizational complexity (e.g., incumbency, communication, dynamicity, sustaining business, social, psychological, and cultural) combined with the complexity of AI

Figure 3.1 Strategic AI considerations.

technologies provides challenges to their adoption. Despite these complexities and associated challenges, AI enhances the ability of the business to respond to changing external and internal circumstances.

Customers, suppliers, and strategic partners are all able to derive benefit from an organization-wide use of AI.

Figure 3.1 summarizes the four aspects of developing a business strategy for data-driven BO. These are as follows:

People

This is the social dimension of BO. Sociocultural factors influencing the business are given high importance. Leadership focuses attention on how BO will affect clients, employees, and other users. For example, changes to work formats (e.g., telecommuting, telemarketing) and their resultant impact on the organizational and social structures are all part of this social dimension. Due consideration needs to be provided to areas of individual and team strengths and the corresponding weaknesses. For example, customer-facing individuals can change the perception of the organization with basic training and a positive attitude. Senior managers and leaders of the organization, working as individuals, can also have a substantial effect in changing to an agile business.

These organizational changes, however, cannot be sudden when people are involved. Training, motivation, and individual aspirations need to be considered, and both performance and functionality need to be kept in balance.

Broader social issues such as the effects of promoting agility, balancing risks with advantages, and ethical and legal business practices (including relevant documentation) are also part of the social dimension. Usability/user experience analysis is also a people issue because of its subjective nature.

Strategic planning for BO starts with a list of key stakeholders (e.g., customers, staff) and the RACI matrix.[10] The level of acceptance by the stakeholders of new ways of doing business and the mechanisms for communicating with them on a regular and consistent basis are crucial to the strategic plan; risks associated with privacy of data and the perception by users is people issue. Training and coaching of users and staff is an important people consideration in BO plans.

Process

This process dimension deals with how the business conducts its transactions both internally and externally. Business process is an integral part of this dimension. AI is considered strategically to change the business processes of the organization to Lean-Agile. This change to the processes impacts the way the business interacts with customers, the way in which it manages its employees, and the way it sets up and conducts collaborations with other business partners. Process consideration in BO enhances BO quality and value provided to the interacting parties without sacrificing the current offerings of the business.

Strategic planning in BO includes which processes to optimize and the extent to which they have to be optimized. For example, in an insurance process, the strategic approach of business leadership can limit the AI-based analytics to only predicting the possibility of a claim without going into the actual decision on whether to accept the insurance proposal or not. "Certain AI tools are likely to transform the boundaries of your business. Prediction machines will change how businesses think about everything, from their capital equipment to their data and people."[11]

Modeling and examination of the activities and tasks of the business processes is the starting point for BO. Embedding analytics in the processes is undertaken only after they are modeled. Estimates of activities and tasks based on metrics indicate how well they are optimized. For example, redundant activities and tasks are eliminated. Due consideration is required in the use of multiple methods within BO because these methods create friction. BO strategies give due consideration to method friction.

Technology

The technical considerations include AI technologies, tools, networks, and devices. AI technologies are the key enablers of business optimization. Technologies include computing machines and IoT devices, software, algorithms, network, security, and systems. The network infrastructure also handles the security protocols. Data-driven decisions in organization's system are made possible through the AI technologies. Internet-based communications protocols, IoT devices, Semantic Web, and mobile and cloud computing are all brought together by enterprise and business architectures.

Data sources, storages, and analytics are supported by the Cloud. Security, privacy, and quality of data are incorporated in the strategic plans for BO.

Money

The money (or economic) aspect of AI adoption deals with costs, profits, and investments. This money aspect of the strategic plan outlines the costs, budgets, and ROI. Competitor risks, the financial impact, the lack of analytics, and potential customer loss are factored into the strategic plan. Costs and benefits of AI adoption are discussed keeping the ROI in mind. Managing the investments, its customer relationships, and its partners all have financial impact on the AI initiative. The success criteria for AI include enhanced customer experience which may not be easy to measure. Strategic plans include costs associated with change management.

STRATEGIC PLANNING FOR BO

Strategies – tactics – operations

Figure 3.2 shows a typical move from strategies via capabilities, projects, and tactics to operations. Digital strategies maintain a long-term outlook ranging from one to three years. Strategies start with the vision and mission of the organization. Strategies provide the basis to build capabilities. A business capability comprises people, process, technology, and resources needed to execute on AI functions. Capabilities are created through projects. Outputs of projects are tactics which are short term. Operations are immediate and ongoing activities of the organization. Strategies and tactics provide the effort to adopt BO with AI. Business operations and processes eventually deliver customer value.

AI is a disruptive technology that demands new capabilities. Big Data along with its volume, variety, and velocity needs data analytics as a capability. Capabilities also provide the means ingest, store, and analyze IoT data. Capability allows business optimization that enhances decision-making during operations. Analytics improve estimates for revenues and associated costs and, thereby, provide internal business value. BO includes resource optimization, financial optimization, critical asset optimization, information optimization, and knowledge optimization. Investment in people skills, processes, and technologies is required to undertake the strategies to operations use of AI.

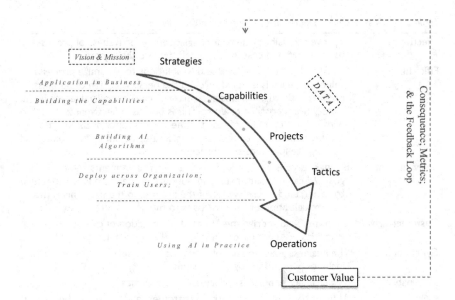

Figure 3.2 **From strategies to operations.**

Digital business capabilities address the key business outcomes. Business capabilities are directly connected to the values, objectives, and goals of the business.

The capabilities of the business are continuously augmented through a rapid feedback loop that is based on the results generated by the initial analytics, corresponding decisions, and their consequences (Figure 3.2).

While data and its corresponding analytics are getting accepted in business decision-making, what is still not fully exploited is storing and analyzing the previous decisions. With self-learning systems, businesses are benefitted not only by the use of data but also by previously made decisions. Decisions based on a large number of previous decisions can be further categorized and used in even faster, more accurate, and automated decisions.

Self-learning systems also need to be self-correcting. This self-correcting nature of our systems requires a feedback loop that is also accurate and quick. Feedback for corrections needs to be provided well within the time required for the next decision.

Capabilities are developed through projects. These AI implementation projects smooth the information flow and optimize decision-making. The people resources are further supported through training and coaching. BO tactical implementation thrives on flexibility in the digital strategy, alignment of capabilities with strategies, and agility in projects.

Digital capabilities also enhance management practices, resourcing, intelligence, and processes to extract value from data. Multiple dimensions of the organization are brought together via digital strategies in order to

Table 3.1 BO strategy characteristics to optimize business processes

Effective	Improve accessibility, timeliness, agility, and effectiveness of processes used by staff and customers to help achieve their goals.
Efficient	Prioritize business functions and AI solutions to maximize process value. Efficiency is enabled through the personalization of business processes through AI. Efficiency may complement effectiveness but it is not always possible to do so. This is because effectiveness deals with the achievement of results, whereas efficiency deals with the most optimum way to those results. Strategies aim to harmonize and personalize functions to enhance efficiency.
Relevant	Business processes should meet the user requirements and business goals. AI can help automate and optimize processes but AI is not able to understand the relevance of the process to the user. This is the contextualization of the processes to the needs of the customer.
Customer-driven	Customer values drive the AI-based optimization of processes. Priorities of AI are based on value to the customer.
Cyber secure	The business goals and AI solutions should be secured. This security includes perceptions by the customers.
Sustainable	Sustainability of business process-based analytics.
Agile	AI helps to respond promptly to customers and partners. The processes and systems to be sufficiently agile to enable change.
Practical	AI solutions should be realistic and achievable. AI is carefully embedded in organizations' systems and information.

enable quality response to the organization's core functions. Capabilities align data activities with the goals of providing customer value.

The operational model is transformed from an "inside out" to an "outside in" model. This means that the organization changes its business processes to make sure that the operational model aligns itself to meeting customer needs – rather than creating or offering business values in such a way that the target customers have to change their requirements to receive what the business offers.

Table 3.1 lists the principles to achieve strategies that enable alignment of outcomes and business capabilities for BO.

ML types in BO strategies

ML in AI is performed at various levels of abstraction (Table 3.2). The rules, patterns, heuristics, and meta-heuristics of ML are summarized in Table 3.2. At the start, domain experts define the rules for the learning algorithm to work on the target data. Definitive rules like "Customers buying product X also buy product Y, Z% of the times," for example, led to data-mined discoveries like "young fathers buying beer cans also buy diapers."

Table 3.2 Machine learning types corresponding to abstraction and task complexity levels

Abstraction level	Task complexity level	ML type
Rules	Find data matching the rules	Data mining
Pattern recognition	Find consistent patterns in a homogeneous dataset	Supervised, unsupervised ML
Heuristics	Find related patterns in a heterogeneous dataset	Transfer learning
Meta-heuristics	Learn how to learn	Self-learning

At the basic level of abstraction, the machine is instructed where to begin its learning, and what kind of patterns to look for. In other words, it is provided with the "what and how" of learning. At the next second level of abstraction, the machine is given a large quantity of data, and through a specific algorithm instructed to look for patterns in the data. The machine is given the how of learning and it finds the what (contents) in the form of patterns. Supervised and unsupervised learning typically work along the algorithmic "hows." The third level of abstraction is applied in transfer learning, where the machine is told what to look for, but has to figure out for itself how to look for what it is looking. In other words, a machine learns the heuristics, just as experts learn the heuristics in their field of expertise after amassing a great depth of expert knowledge. This is the deepest (or the highest) level of ML that is currently available. Arguably, these three levels of abstraction in learning are not adequate to respond to all business needs. A future level of abstraction where machines will learn how to learn and what to learn is needed. They will be provided neither with the what nor the how of learning. Rather, based on the lower levels of abstractions, they will arrive at the level of meta-heuristics, where they will engage in learning the very art of learning, given a context.

LEADERSHIP IN BUSINESS OPTIMIZATION

Leaders hold a strategic outlook that goes beyond the immediate operational and tactical needs of the organization. Leaders envision the future and contemplate its challenges for the business. Therefore, leaders in BO do not limit the vision to AI technologies but go beyond into the customer value, user experience, user perception, and related subjective spaces.

Anticipating the future includes the creation of multiple "what if" scenarios to handle the business response as the reality changes. Good leaders carry the many "what if" scenarios in their heads all the time. "Technology, however, is dramatically rewriting the rules of business and life."[12] For example, ML algorithms self-propagate themselves – requiring a deeper understanding of the need to "explain" the insights generated. The absence of "explainability" of AI can create sociocultural challenges that are way beyond technologies. Another example is the decision on the level of granularity of analytics. The extent to which fine-granular analytics are required is also a strategic question based on the need of the business.

Given the explosion of "data," AI tools and technologies are required to create multiple futuristic scenarios. Data, combined with experience, is the fuel for the decision-making engine of leadership. Leaders also look beyond the traditional sources of data and delve into using alternative data. Leaders also ensure that automation with AI is complimented with optimization and humanization of business processes.

Automation strategies

Automation repeats the same business function and process – albeit using technologies. As a result, automation makes the same process faster. Since the process is coded and tested, the chances of error in an automated process are greatly reduced. Automation strategies may not incorporate the creativity and human ingenuity. Therefore, Natural Intelligence (NI) is not a part of automation strategies.

Automation strategies are important in the business optimization because of the following reasons:

- Mundane tasks that people will not enjoy can be offloaded to machines. Robotics is an example of automation wherein AI-based machines control operational processes.
- Routine and well-defined tasks are performed much better (faster and with greater accuracy) by machines. For example, inventory management and SCM.
- Machines can automate large processes in parallel thereby further reducing the time required to perform the tasks.
- Automation provides ability to provide detailed data-driven audit trails that can be used for analytics including forensics.
- Machines can measure, report, and improve their performance.
- Artificial intelligence fast tracks actions as time is not lost in determining the cause – it is already known.
- Automation also supports business continuity.
- Machines learn from each episode and thereby improve their predictability.
- AI systems learn from previous disruptions, predict the use of alternate resources, and recommend automated response and recovery options.

Optimization strategies

AI is strategically used to optimize business processes. This optimization starts within the BO strategy by assessment and prioritization of existing processes. Strategies enable AI systems to be as self-organized and optimized as possible. AI takes BO to the next level. Optimized processes are able to understand the urgency of a customer need and the severity of an issue, and identify approach to the solutions. The AI-based optimization of business functions makes use of people, processes, money, and technology.

Following are some key aspects of optimization strategies:

- Business are made up of multiple attributes. Each attribute has multiple perspectives depending on the context. These attributes, or parameters, are identified, documented, and customized based on the business goals.
- Optimization evaluates the organization's technical capabilities. Assessing current performance levels (including gaps, risks, inefficiencies, and opportunities) against required performance provides opportunities and direction for the application of AI to business processes.
- Optimization aligns strategic intents and outcomes with capability-building initiatives in the business. Continuous alignment of strategic intents with capabilities and projects is essential for BO.
- Optimization uses capability management to provide resources and their allocation to meet various competitive dimension.[13]
- Optimization of business needs governance of organizational resources to enable focus on the desired business outcome.
- Optimization strategies focus on enabling the users to achieve their objectives in an effective and efficient manner. Time and accuracy are treated as enabling factors in optimization strategy.
- Optimization strategies require a change of the mindset of the organization to a customer-focused one.
- Optimization strategies develop models for data assets and corresponding processes keeping the end-user in mind.
- Optimized processes implement policies and rules electronically to enable efficiency.
- Cybersecurity ensures the safety and integrity of customer data.
- Optimization strategies assimilate disparate and disjointed processes in order to collaborate and eliminate redundancies.
- Optimization models existing processes and policy frameworks to define gaps. Processes help understand the impact of adoption of AI better.
- Optimization blends the physical, electronic, mobile, and collaborative processes to provide a unified view of the business to the customer.
- Creating a unified view for the customer (through MDM) also enables a single point of contact for the customer.
- Optimization aims to provide insights for decision-making at the point of action.

Humanization strategies

Humanization importance is understood by investigating the nature of AI models – which, by its very nature, is performance driven. The users and the business leaders aim to maximize the performance. ML code embedded in AI models is agnostic to the specific situation of the user. AI systems are based on data correlation, whereas human systems are based on cognitive knowledge and understanding causation.

AI systems are not cognizant of human values on their own. As a result, business decisions can result in a loss of value. Due consideration to this loss of value is a crucial element of "humanization." Providing customer value is a subjective phenomenon. AI cannot understand or produce value on its own. The humanization of AI-based decision-making is a strategic imperative. AI-based predictions and user experience is the right judicious combination needed for decision-making.

The application of AI together with NI for prediction, handling of the crises, and recovery of business is part of the humanization aspect of the BO strategy. Humanization ensures that AI-based decision-making does not replace people entirely. Instead, the human factor is brought in the automation and optimization of business processes in a balanced manner.

Intellectual property, people skills, processes, and finances are crucial organizational resources that are put together in humanizing strategy. Customer engagement and operational management work together. For example, for a given customer engagement, if the tasking and coordination systems are not integrated with human resource management systems, resource allocation is manual. Such resource allocation has duplication and results in multiple handling of the customer queries.

Sourcing people (recruiting) and maintaining skill levels is part of human capability building. Empowering team members to take quick and innovative decisions and provide high-quality services requires a change of mindset. This change of mindset also requires a corresponding change in the organizational structure that will facilitate and nurture the decision-making capabilities of staff at the customer-facing level of the organization.

Mindset changes in style, structure, culture, and skills of people. A flexible and agile organization structure reflects changing context in which the business operates. The need to empower staff members and encourage them to use AI-based decision-making is a humanization strategy. Upskilling people is made up of sourcing new people and training existing people that result in a workable mix of business, technology, and functional capabilities.

Users and culture changes

The AI-based culture shift of the organization involves a number of structural, social, and individual changes. Users need the opportunity to try new approaches and learn lessons. Developing soft skill contributions of the team members and managing team spirit are agile leadership characteristics. Agile also relieves stress, facilitates communication, and enables team spirit. The hierarchy reporting structure changes to a flat one. Individual personalities and their biases come into play in humanization.

Visibility, collaboration, discussions, and objective reality of the environment in which the business operates is beneficial in developing a digital business strategy. Cross-functional teams at the business, corporate, and departmental levels work to iteratively define and manage these strategies without the defined timeframe on an ongoing basis. Each strategy outcome is mapped to the business capabilities and the desired business outcomes.

BUSINESS OPTIMIZATION INITIATIVES

Business optimization is based on optimization of business processes. AI/ML helps in identifying redundancies, gaps, and lags in processes. The starting point for a strategic approach to BO is the desired business outcome. Understanding the desired outcomes provides the basis for the BO initiative. AI-enabled BO is based on data. Therefore, understanding of the outcome is followed by a detailed understanding of the data, including its sources, quality, and costs. Approach to analytics and the mechanisms to embed the results in business processes are outlined next. Eventually, the analytics are presented – mainly through visualizations. Visualizations include a wide range of techniques, including bar and pie charts, bubble charts and heat maps, and so on. They show trends and patterns in data.

Figure 3.3 shows the organization factors impacting the BO initiatives and their role in developing the BO strategies. Business context provides the business objectives and key business requirements for the AI and Big Data initiative. The two subfactors arising from the business context are the financial impact and ROI of the entire initiative and the GRC. The financial impact and ROI of the BO initiative leads to an understanding of how the AI capabilities can be applied to understand the markets. The opportunity costs associated with the initiative and the possible loss of opportunities without AI are established. Understanding the technical and business capabilities of cybersecurity and disruptions and recovering from them are outlined next. Data, analytics, networks, devices, processes, and people are each explored in terms of how they contribute to the BO initiative. Each aforementioned element has a strength and a risk – which are explored and documented in the context of the particular business.

Strategic approach to BO initiatives covers the entire data lifecycle – starting with data input, then understanding the context of that data, establishing the controls for the inputs and their variations, deciding on the level of granularity, and creating a feedback loop based on the results. This data lifecycle stands to benefit by an agile approach to developing analytics. For

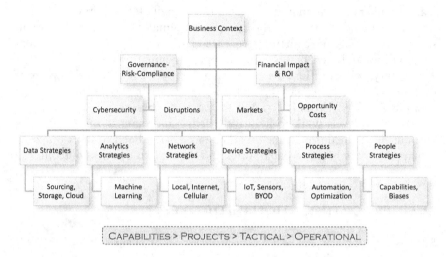

Figure 3.3 Organizational factors impacting BO initiatives.

example, not all of the above factors in the data lifecycle need to be available at the beginning of the development. These factors can be iteratively provided as the solutions are developed and presented to the users. Agility in solution space supports an iterative, incremental, and collaborative approach. Agile approach to solutions includes creation, evaluation, and continuous refinement of analytical algorithms. Data scientists continue to interact with analysts from statistic and business process space for further explorations, analysis, and profiling. A functioning and validated analytical model can then be applied to new, incoming data suites.

Developing a business case for AI in business optimization

Developing a business case for the use of AI in optimizing the business is based around the value proposition to the customer. AI is considered as an indispensable tool for automation and optimization. The business case, however, is based on how it enables the business to provide enhanced customer value. Earlier discussion on business strategies translates to a strategic plan. Such strategic plan for BO includes the costs associated with AI and the value generated from it. The strategic plan also includes the risks associated with the changes in the operating model of the business.

The strategic plan includes the details of the strategic approach, the gaps and, therefore, the needs to develop new AI-based capabilities, the projects needed to develop the capabilities, and the eventual value to the customer when the solution is deployed in operations.

Specifically, the business case and justification for AI adoption in business contains the following:

- An executive summary that contains the key numbers in terms of costs and returns, as well as risks associated with the adoption of AI, competition, and GRC.
- Description of the problem and opportunity due to AI. This is the SWOT analysis of the organization.
- The business goal of optimization – specifically in terms of providing customer value?
- Key stakeholders in the BO initiative – including the staff, the data scientists (and related roles), collaborating partners, customers, and regulators.
- The time and budget parameters of the organization for the AI initiative?
- The risks associated with the AI adoption and the risks of not adopting AI.
- Change management approach for the business processes based on automation, optimization, and humanization?
- Description of the resources for the initiative. Will the people resourced be developed internally or sourced from outside?
- Underlying issues within the data and meta-data such as its quality, ownership, security, and privacy.
- Key issues in implementing the algorithms and using the data in the business processes.

The following "Ps" are an ideal checklist for developing the plan:

- Planning – who will plan, who will be impacted, and for how long?
- Potentiality – of various AI capabilities in the context of business.
- Proactivity – from the business leaders in developing the BO strategy and handling change.
- Prediction – capabilities of AI and their technology implementation challenges within the business processes.
- Prescriptive – What are the prescriptive capabilities of AI that can help the business handle disruptions and ameliorate crises?
- Prevention – How can strategies be developed using AI that will help the business anticipate, prepare for, and, if possible, prevent disruption?
- Protection – How will the business be protected from cybersecurity threats?
- Performance – How to ensure the AI solutions in operation does not suffer performance degradation due to heavy analytics and/or security of the process?
- People – changing the mindset, training, and mentoring. Who (what roles) are involved in decision-making for this business area?

Business stakeholders in strategy

Developing AI strategic plans requires due consideration to time, budgets, resources, risks, agility, metrics, people, and adoption. People are the stakeholders, mentioned in the strategic planning documentation above, who make or break the BO initiative in an organization. Therefore, all stakeholders need special attention in developing digital business strategies with AI and Big Data technologies. The leaders are involved in developing these strategies in consultation with the other business stakeholders. The business stakeholders ensure the start of the initiative, budget for it, and aim to provide the value to the customer; the technology and project stakeholders are interested in providing an efficient and effective AI solution.

Stakeholders can be responsible, accountable, consulted, or informed. In practice, though, their roles and responsibilities change. Furthermore, one person may perform multiple staff roles – for example, a doctor in a hospital also performs some administrative tasks. A customer can also change roles. For example, a patient getting admitted to a hospital is also making payment for a health insurance. Modeling the requirements of a process and then embedding analytics within them is given due importance in developing the strategic plan. Agility in business and in developing, configuring, and deploying AI solutions enables iterations and increments – so crucial to the digital business strategy.

Following is a brief description of some of these stakeholders.

- Business leaders and other business people – these are the initiators and "adopters" of AI and Big Data in the business space. These stakeholders include the business sponsor, the investors, and the domain expert. Incorporating artificial intelligence into decision-making is undertaken keeping the "humanization" aspect in the strategies. This aspect requires the leaders to communicate regularly with the rest of the organization, including staff, managers, and also external partners and customers. Leaders provide the support during change management.
- Technology leaders, including the Data Scientist, Data Analyst, and other technical and project roles that will be responsible for delivering the solution.
- External (partners) stakeholders are the collaborators from other business partners who are jointly offering products and services. Collaborators can also be suppliers of data, especially alternative data that is used in the AI solution.
  - Users of business processes – These are typically the staff of the organization who are using the AI solutions. These users make use of the analytics within their processes to serve the customer. The users can also be senior business people who provide input in the direction of a product.

- Customers who consume the product or service being offered. Customers also use the analytical outputs to self-serve themselves as well as provide opinions and feedbacks to the business. Customer representatives also provide requirements for the AI-enabled business processes.
- Legal and compliance stakeholders – can be both internal and external. Internal roles include legal and audits that ensure compliance as AI is embedded within the business processes, and external are the regulators who monitor the compliance typically from security and privacy angles.

STRATEGY CONSIDERATIONS BEYOND AI TECHNOLOGIES

Strategies to incorporate natural intelligence (NI)

NI is not linear. While humans break up large and complex problems into smaller bits to be able to better focus, the eventual solutions are holistic. This is particularly true with AI solutions – and therefore, they need NI. Comprehensive AI systems learn and enable learning using multiple points and in a nonsequential manner.

Handling business problems requires a combination of AI and NI. AI algorithms on their own may not properly understand the scope and context in which the solution is operating. Even if humans and contemporary machines had the ability to analyze the data, the demand for increasingly higher speed to analyze the data and the influx of high-velocity data implies a judicious combination of AI with NI.

Transfer learning which is discussed at length in the context of dynamicity of learning in Chapter 5 is an attempt to include the paradigm of learning from multiple points nonsequentially. As a result, deep learning (DL) models that are trained on limited datasets need not be redesigned and reconstructed to learn from totally new, but related datasets. These models can transfer their inner knowledge and adapt themselves to learn from other datasets and circumstances. Although still at an experimental state, DL models are gradually picking up steam when exposed to Big and multidimensional data, thus reducing learning times by orders of magnitude.

Strategies for formulating the problem

An important aspect of strategic planning for AI is that of formulating questions for the solution. What are the kind of questions that an AI solution can offer? And what should be the approach to capitalize on the insights, the answers? Interestingly, the underlying Big Data on which AI algorithms work is so vast and so complex that not only are businesses not able to ask the questions of that data, they fail to recognize what those questions should be. Strategic plans incorporate approaches to not only find the solutions but also how to approach the data to ask the right questions.

Strategies for formulating problems necessarily require NI. Automated planning, automated prediction, and automated decision-making are AI-driven to make business processes efficient and agile. But the reverse process of formulating problems or asking questions to improve business processes is still beyond current AI technology. Optimizing single functions at a time is often enough to solve quite a few problems. However, most practical and business problems are very complex because they present scenarios that call for simultaneously optimizing diverse functions, which are often conflicting in nature. AI is used iteratively to provide partial insights which are used to formulate the problems themselves.

AI is also used iteratively to make sense of data that no longer limited to text-based data, nor is it able to be organized in traditional rows and columns. Data variety includes free-flowing text, blogs and reports, emails and messages and tweets, and even photographs and videos. These varieties of data require AI strategies that incrementally assimilate different types of data, their aggregation, and subsequent analytics on that data.

Multiobjective optimization is a well-developed discipline containing advanced and efficient algorithms that can handle a wide variety of conflicting objective functions, as well as constraints. However, the design and choice of functions require excellence and experience in data science as well as the problem domain.

Strategies for improving quality of decisions

The more iterative and incremental approach in the AI implementation, the greater is the quality of the solution. While not all AI implementations are iterative and incremental, strategic planning for BO can consciously insert iterations and increments.

Incorporating the impact of the decisions into a "decision engine" improves the quality of decisions. The quality and impact of decisions are evaluated by the AI engine resulting in a rich source of input into the overall decision-making by leaders. These capabilities of anticipating and predicting disruptions are continuously improved – on an *ongoing basis (Kaizen* style).

AI-based systems need to anticipate the quality challenges and recommend corrective actions. Sustainable and continuous business operation is based on AI-based recoveries. This sustainability and reliability are a crucial value-add to the customer. AI-based business strategies include improving the quality of decisions and value to customers.

AI AND BUSINESS DISRUPTIONS

Disruptions due to AI as part of strategic planning

Businesses are based on various types of processes that are external facing (e.g., reaching out to the customers) and internal facing (e.g., inventory, HR). Applying AI to routine or known business is automation. These are the processes that undergo change and are disrupted when AI-based optimization is brought about. Implementing AI to optimize business processes is not a separate, standalone initiative. In fact, BO has to occur mostly when the current business processes are in operation and the business is actively operating. Decision-makers need to learn to use their knowledge, expertise, and experience together with what the analytics are suggesting. Similarly, customers need to learn to self-serve themselves whenever possible.

Strategic planning considers the disruptions to these processes when AI is implemented.

Incorporating AI to handle externally imposed disruptions to business

AI has increasing accuracy in predicting as it "learns" from its previous predictions. AI, therefore, can predict disasters and automate business continuity. Leaders can proactively monitor challenges as they are alerted by AI systems.

Disaster recovery strategies benefit the most with the use of AI because AI enables making sense to data that human capabilities fall short of. Data, coming and going out of an organization through myriad channels, provides the strategic base for disaster recovery and business continuity.

For example, leadership at both technical and business level was heavily challenged during the COVID-19 pandemic. Businesses around the world were severely affected with many closing down, jobs being lost, and the global economy on the brink of depression. Could businesses have been better prepared to anticipate and act on this disaster? Could data and corresponding AI tools have made any difference to the way businesses responded to the pandemic? Could decisions based on previous major disruptions have provided sensible decision-making in response to the pandemic? Although the pandemic provides the greatest opportunity to explore, experiment, validate, and act on the responsibilities of leadership, these questions are not just limited to the pandemic.

AI plays a phenomenal role by creating data models and predictions on use of resources, anticipation of peaks, optimization of supply chains, and management of disrupted businesses processes[14] (e.g. during lockdowns through the pandemic). Codifying the experience of a disruption and the knowledge gained during the response is also an important activity within AI. Disruptions create chaos in business and society. These disruptions can also be an opportunity for innovation and change. Leaders in business are entrusted with the task of holding a strategic outlook. Business decisions can positively impact innovation and change in business if the disruptions are predicted earlier in time.[15] AI systems can auto-launch appropriate actions that are complemented by human decisions. Humans can initiate a response with support from AI. Large-scale social disruptions, such as the COVID-19 pandemic, or the Beiruit ammonia explosion, are examples where AI systems help initiate auto-response.

COVID-19 has demonstrated the need for businesses to anticipate and act on "unknown-unknown" disruptions. Authors have discussed practical, emotional, and even spiritual strengths needed within an organization for resilience in a time of crisis.[16] Disruption categories are five specific groups as summarized in Table 3.3.

Business disruption prediction framework (BDPF)

AI-based business disruption prediction framework (BDPF) helps with anticipation and action around a crisis.

The BDPF shown in Figure 3.4 is made up of two distinct DL engines: DLE1 and DLE2. Data is sourced from within and outside the organization (making it a Big Data storage). The historical, sensor, crowd, and third-party data, after passing through an adequate preprocessing stage, feeds into the 1st stage DLE1. This engine spots the anomalies in the data and learns to classify the potential disruptions into five different categories as per Table 3.3: technology breakdown, cybersecurity breaches, natural catastrophes, societal disruptions, and others (unknowns). For example, DLE1 learns to predict technology breakdown from sensor data, security breaches from server logs, natural catastrophes from weather reports, and societal disruptions from web sources and social media, and places them in an organized manner within a decision engine. Virtual machines are used for a cloud-based auto recovery. Optimizing the recovery includes time and accuracy factors.

DLE2 is a transfer learning clone of DLE1, which means that it has all the learning experience accumulated by DLE1 and can make finer predictions based on the currently available online data. It also has three additional inputs indicated by arrows: (a) Prediction-Experience information supplied by decision-makers; (b) what-if-analysis data input; and (c) feedback loop from the decisions made by the management based on the predictions of DLE1 and their past experience and expertise. These three inputs further refine the disruption prediction. Since the BDPF continuously monitors the ever-changing data in the online data sources and automatically feeds the processed data to the DLEs, the predictions due to the potential future disruptions are made in real time. This real-time processing of data provides early warnings for the business leaders to decide and act to mitigate the disruptions.

Table 3.3 Role of prediction models (AI) in anticipating disruption

Disruption categories	Description	Prediction	Experience
Technology breakdowns	Systems, databases, Processes and devices breakdown – especially at crucial junctures in the technology ecosystem. Administration and production will not function	Data is sourced from IoT devices, third-party Cloud-based services, previous historical data stored in the organization's own databases	Experienced Enterprise architects and business architects support business leaders in anticipating breakdowns
Cybersecurity breaches	Data and network breaches leading to loss of privacy and security	User logs and access points are analyzed in order to highlight the anomalies in the function in advance	Experienced cybersecurity experts and enterprise architects help business leaders create various "what-if" scenarios
Natural catastrophes -	Earthquakes, tsunamis, volcanoes, hurricanes, landslides, epidemics, and pandemics. The catastrophes do not have a hidden agenda. Production and supply chains will halt; staff will not report to work	Data is sourced from news bulletins; seasonal changes in data; AI engine (analytics) provides early warning of the catastrophe based on previous occurrences and corresponding similarities in parameters	Ability to spot the changes in data-driven visuals; ability of leaders to put the disruption management strategies in action as soon as possible after the event
Societal catastrophes	Admin, production, and supply chains will halt, and staff will not report to work; legal changes, wars, strikes, demos, and protests. These can be vicious catastrophes with long-lasting impact because of their clandestine and ulterior nature	Social media data (Tweets); crowd data; various discussion groups, blogs, and news items; web scanners can scrape data off from sites and apps in order to build trends	Previous experiences of societal disruptions enable leaders to make an educated guess in terms of the coming societal disaster; Experience also helps in planning for and swinging into action when the event occurs
Others (uncategorized)	Combination of natural and manmade factors	Data across multiple public domains is monitored for anomalies. Virus occurrence (hospitalization data), its spread path, and actions to take	Leadership experience of business, industry, society, and the overall environment is most helpful in anticipating and handling unknown disruptions

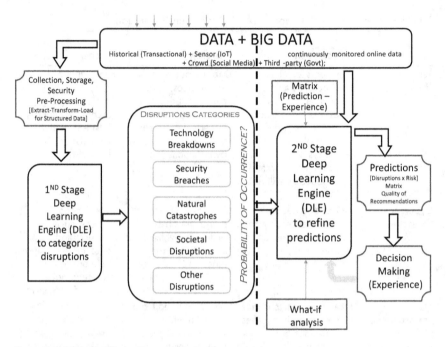

Figure 3.4 Business disruption prediction framework.

The what-if analysis input is provided to DLE2 to make timely and fine-grained predictions. This is so because the what-if analysis creates an opportunity for iterative and incremental improvements in anticipating disruptions and taking actions. The BDPF shown in Figure 3.4 is not focused on automating the disaster recovery of an organization. Instead, it is based on a judicious combination of AI and NI through a healthy interaction or collaboration of machines and humans.[17] Such AI-NI balance enables the leadership to bear their experience on the predictions generated from the AI engine depending on the type of disruption.

BDPF is also applicable in the present to the COVID-19 pandemic, which demands numerous decisions to be made based on the anticipated rate of growth of the disease, necessary treatments, and so on. This pandemic which falls in the unknown-unknown category (as discussed in Chapter 2, Figure 2.6) could benefit immensely through intuition. The intuitive approach, however, is not the approach a business can adopt as its formal anticipation and action towards disruptions.

The BDPF creates the opportunity for the establishment of a correlation between existing organizational data, external higher-level (governmental) data, and the corresponding policies and guidelines. Predictions, backed by experience, is the best opportunity to move the handling of COVID-19 pandemic from the unknown-unknown to the known-unknown or unknown-known quadrant of Figure 2.6 (Chapter 2). AI superimposed with NI results in better decisions such as the early release of lockdown in Japan during the COVID-19 pandemic.[18] Agility in the BDPF enables the creation of numerous scenarios and prototyping of the responses of the leaders and their businesses. The BDPF makes it easier to enable a rapid feedback loop on the business level decisions. The accuracy of decisions can also be improved.

The BDPF also identifies the *shifts* in disruptions (possibilities) to ensure that its predictions are not mono-dimensional. The decision-making part includes DL algorithms from BDPF which have the ability to accept "impact of decisions" including their weighting and biases in the ongoing analytics and predictions for future decisions. BDPF identifies potential for disruptions and flags them in an appropriate manner to enable actions by decision-makers in business. The results of the actions continue to be fed back, with appropriate weighting and risks (as judged by business leaders), into the ML engine on a *dynamic* basis. This ML decision engine framework further ensures the capability of machines is put to use to identify improvements in predictions.

CONSOLIDATION WORKSHOP

1. What are the key strategic AI considerations for a business as it aims for BO?
2. What are the important nuances of leadership in the digital era? How do they differ from a non-digital era? What should the leaders do in the initial phases of BO?
3. What are the justification mechanisms for the use of AI/ML/BD initiatives in an organization? How would you incorporate them in a strategic plan?
4. How are digital capabilities defined in BO? And how are these capabilities aligned with strategies?
5. How are business capabilities required to meet the strategic goals of the organization be identified?
6. What are the risks and rewards of digital technology on a business?
7. What is the difference between automation, optimization, and humanization?
8. How can constant disruption be anticipated and how are strategies developed to use data-driven decisions to handle the disruptions?
9. What reactive and proactive approaches are developed to handle the digital technology disruptions?
10. How do strategies incorporate risks and how are those strategies managed?
11. How can an organization plan for projects that will develop digital capabilities?

12. Why is it important to position tactics and operations in digital business strategies?
13. How can long-term and short-term gains be balanced by the application of AI/ML?
14. What are the efficiency and effectiveness of gains (ROI) using AI/ML?
15. Why is it important to define resources and timelines of business strategies?
16. How can leaders strategize to manage change due to disruptions?
17. What is the Business Disruption Prediction System (BDPF)/Meta-Model for Decision-Making?
18. How can a BDPF ensure successful application of AI / ML to business optimization?
19. How can actionable insights be developed to enhance organizational capabilities?
20. What are some ways to integrate data and processes with decision-making?

NOTES

1. Tiwary, A., and Unhelkar, B., (2018), *Outcome Driven Business Architecture*, (CRC Press, Taylor & Francis Group /an Auerbach Book, Boca Raton, FL).
2. Unhelkar, B., (2013), *The Art of Agile Practice: A Composite Approach for Projects and Organizations*, (CRC Press, Taylor & Francis Group /an Auerbach Book, Boca Raton, FL). Authored ISBN 9781439851180, Foreword Steve Blais, USA.
3. See note 2.
4. Unhelkar, B., (2018), *Big Data Strategies for Agile Business*, (CRC Press, Taylor & Francis Group/an Auerbach Book, Boca Raton, FL). ISBN: 978-1-498-72438-8.
5. See note 2.
6. Tiwary, A., and Unhelkar, B., (2018), *Outcome Driven Business Architecture*, (CRC Press, Taylor & Francis Group /an Auerbach Book, Boca Raton, FL).
   Hazra, T., and Unhelkar, B., (2020), *Enterprise Architecture for Digital Business: Integrated Transformation Strategies*, (CRC Press, Taylor & Francis Group Boca Raton, FL).
7. See note 4.
8. See note 6.
9. See note 2.
10. https://en.wikipedia.org/wiki/Responsibility_assignment_matrix accessed 1 Nov 2020.
11. Agarwal, A., Gans, J., and Goldfarb, A., (2018), *Prediction Machines: The Simple Economics of Artificial Intelligence*, (Harvard Business Review Press, Boston, MA).
12. Downes, L., (2009), *The Laws of Disruption: Harnessing the New Forces That Govern Life and Business in the Digital Age*, (Basic Books New York, NY).

13. Hammer, M., and Champy, J., (2001), *Reengineering the Corporation: A Manifesto for Business Revolution*, (Nicholas Brealey, London).

14. Gonsalves, T., (2017), *Artificial Intelligence: A Non-Technical Introduction*, (Sophia University Press, Tokyo).

15. Gutsche, J., (2020), *Create the Future + the Innovation Handbook: Tactics for Disruptive Thinking*, (Fast Company PR New York, NY).

16. Tsipursky, G., (2020), *Resilience: Adapt and Plan for the New Abnormal of the COVID-19 Coronavirus* Pandemic Kindle Edition, (Changemakers Books United Kingdom).

17. Unhelkar, V. V., Li, S., and Shah, J. A. 2019, "Semi-Supervised Learning of Decision-Making Models for Human-Robot Collaboration." In *the Conference on Robot Learning* (CoRL), Osaka, Japan.

18. https://time.com/5842139/japan-beat-coronavirus-testing-lockdowns/ (retrieved on May 24, 2020).

Chapter 4

Machine learning types

Statistical understanding in the business context

MACHINE LEARNING OVERVIEW

Traditionally, solving problems using a computer involves writing detailed instructions in the form of a code. Machine learning (ML) extends this problem-solving ability of computers without being explicitly programmed. Alan Turing, in a talk given to the London Mathematical Society in 1947,[1] predicted ML, saying "what we want is a machine that can learn from experience." Later, in 1959, Arthur Samuel defined ML as "the field of study that gives computers the ability to learn without being explicitly programmed."[2] Tom Michel gave a formal definition of ML as, "Machine learning is the study of computer algorithms that allow computer programs to automatically improve through experience."[3]

ML is of interest to business because of its ability to solve business problems. This discussion on BO is a business issue that aims to use ML to enable it to provide customer value.

Traditionally, computers do exactly what they are told to do. *Algorithm* refers to a detailed set of unambiguous steps given to the computer to solve a problem. These steps are coded in the form of a computer program, loaded in the RAM, and executed. After execution, the results are presented as visuals. Business people use these visuals to make decisions.

Applying ML

Most problems in science, engineering, economics, and finance are solved by means of equations. An equation is generally in a parametric form, where the parameters (or variables) are related to one another. Plugging in the values of known variables and performing well-defined mathematical operations on the equation yield the values of unknown variables. The parametric equations are in a functional form, where the unknown variable is expressed as a function of the known variables. The problem-solving strength of the above disciplines rests on the functions which relate the unknown variable to the known variables. Difficulties arise in solving problems which have no function connecting the dependent and independent variables. ML handles these types of problems.

Data contains a set of variables whose relationships are not established. ML algorithms work on the data to establish relationships between the input (known variables) and the output (unknown variables). In short, given a sufficient amount of data, ML algorithms solve a problem in lieu of the functional form.

ML differs from traditional computing in the sense that the step-by-step detailed instructions are not provided by programmers to solve a problem. Instead, the software program is provided just enough instructions that allow it to learn from data. Machines learn by discovering patterns inherent in data. This data-based learning technique gives the machine power to predict future events/phenomena from fresh domain data and solve a host of problems such as image recognition/generation, speech recognition and synthesis, language understanding and translation, handwriting recognition, medical diagnosis, weather forecasting, game playing, stock prices prediction, customer recommendation systems, fraud detection, self-driving cars, and so on for which there are no human-designed explicit algorithms.

Machine learning steps

ML initiatives in BO start by understanding the business context goal. The organization adopting AI is expressed in its business strategies (Chapter 3). The team entrusted with an ML initiative breaks the problem down into smaller projects in order to handle its complexity. The fundamental steps in developing an ML solution, as shown in Figure 4.1, are as follows:

- Problem formulation – Identification of the overall business problem. Based on the complexity of the problem, break it down into smaller, understandable problems, just as breaking down high-level processes or business functions leads to better handling for optimization. This breaking down of the problem (or decomposition) is followed by framing the problem in ML format. Some parts of the problem could contain labeled data, others unlabeled data, and so on. Breaking down the problem helps apply an appropriate ML-type algorithm to solve the problem.
- Data preprocessing – Identify the necessary data, its source, and its location. Undertake preprocessing of data to ensure it is clean. Cleanliness of data includes the quality of its contents and also of its formatting. Sources and location of data are important considerations in this step.
- Model building – Select an appropriate ML algorithm and build the solution model, keeping the business context in mind. There will be some iterations between this and the previous step as, in the first instance,

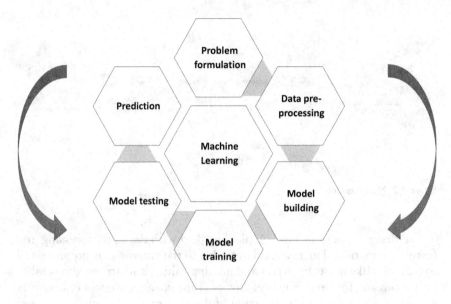

Figure 4.1 Steps to solve a business problem via ML.

model and corresponding data may not match. Refinement of ML model and/or reformatting of data is expected in the first, Agile iteration.

- Model training – Development of a training dataset to be used to train the model developed in the previous step. ML model is then executed on this dataset to enable it to start "learning" from the data. Typically, a training dataset can be about 70% of the overall data cleansed in the earlier step.
- Model testing – This step is the verification of the trained ML model in terms of its accuracy. Testing data can typically range from 10% to 30% of the prepared data.
- Prediction – Embedding the model in a business process and upskilling the users to start using the analytical results in order to automate/-optimize their processes.
- Maintenance – This is an ongoing step of maintaining the currency of the ML model for its accuracy and relevance to the business process in which it is embedded. Should the source, format, and velocity of the incoming data change, this step will ensure that the ML model is ready to handle that change.

Each of the steps mentioned above involves multiple substeps which are separately designed. Unifying the major steps in ML in a framework makes the coding task easier. Often, the steps are automated in what has come to be called an ML pipeline as shown in Figure 4.2.

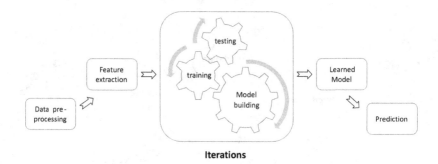

Figure 4.2 ML pipeline.

The first two steps in the ML pipeline deal with data preprocessing and feature extraction. The raw data available in the enterprise is preprocessed to fix issues like noise, incorrect and missing values, and arranged in a tabular form ready for further processing. Feature extraction refers to carefully weighing, selecting, and rearranging of data characteristics and properties by domain experts as the outcome of ML heavily depends on the input data features. The individual stages of model building, training, and testing are commonly integrated in an ML implementation framework.

Model building refers to selecting an abstract static construct like linear regression, support vector machines, decision tree, neural network, and so on, and determining the exact building blocks and the accompanying parameters that make up the model. The model constructed with the available data in mind is then dynamically trained and tested with training and test data subsets, respectively. The outcome of the integrated building, training, and testing steps is a learned model. The final step in the ML pipeline is the monitoring and use of the learned model for making predictions on new and fresh data. These steps are executed in an iterative and incremental manner following the Agile approach to delivering software solutions.

ML TERMINOLOGY

This section briefly describes some of the frequently occurring terms in ML literature. These terms deepen the understanding of ML models and learning processes explained later in this chapter.

Model

Models are mathematical or abstract representations of a construct used to carry on learning. The construct or structure usually is made up of individual components. Unless otherwise stated, the view adapted in this book is that models are generic and static. They are activated by running the algorithm with data on. For example, linear regression equation is a mathematical model for learning linear regression. Decision tree model used in supervised learning is a static tree-like structure made up of nodes and branches. Classification and regression trees (CART)[4] is an algorithm used to train a decision tree using the training data. Similarly, neural networks used in shallow as well as deep learning are made up of several component layers, each containing numerous individual units called neurons. The entire neural network structure is static before being trained by the backpropagation algorithm and training dataset.

Models used in ML could be *parametric or nonparametric.*

Parametric

In the parametric approach, the data scientists, after examining the data, assume a functional form of relationship between the input and output variables. The relationship may be linear, polynomial, exponential, and so on. The data scientists can accordingly frame curve-fitting equations and perform elaborate computations by plugging in the available data to determine the coefficients in the equations. The statistical techniques of determining the coefficients of the equations that represent the functional form between the input and output variables in the dataset are called learning the parameters. Once all the coefficients in the equations are known, the parametric model can be used as a predictor (another name for the learned model) when new and fresh data is acquired.

Nonparametric

In the nonparametric approach, the functional form of the relationship between the input and the output form is not known. Rather, the functional form is assumed to be implicit, and the nonparametric model is trained to extract the pattern from the data and estimates the functional form in an abstract manner. The models are largely general purpose and have wide applications.

Model parameters

These are variables internal to the model, whose appropriate values are learned from the data during the training process. The weights on neural connections (which are a rather poor imitation of the synapses of biological neurons) are typical model parameters. Training the model consists in optimizing these distributed weights so that they predict the expected output for a given input from the dataset.

Hyperparameters

These variables are external to the model and cannot be learnt from the data. They are like control parameters of the learning model. Choosing an appropriate set of hyperparameters is both crucial in terms of model accuracy and computationally challenging.[5] Their values are tweaked or tuned based on some heuristics or programmer's experience. Well-tuned hyperparameters improve prediction accuracy of the learned model. Learning rate and mini-batch size are very important hyperparameters. Inappropriate value of these parameters can lead to model divergence and incorrect prediction results.

Training

The subset of the original dataset called the *training data* (Figure 4.3) is run through the static model. This execution follows a specific algorithm and is repeated a large number of times. In each iteration, the model parameters are systematically tweaked, and the iterations are continued till convergence.

Validation

It is an additional step often used towards the end of the training epochs to get an early estimate of the performance of the model.[6] Validation, which is performed using the validation data subset, is usually performed to tune the hyperparameters.

Testing

The trained model is then fed with the test data and its performance is evaluated. The performance of the trained/learned model should be tested on a held-out dataset that has not been used prior, either for training the model or for tuning the model parameters.[7]

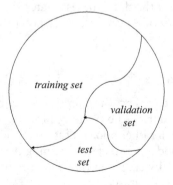

Figure 4.3 Training, validation, and testing data subsets.

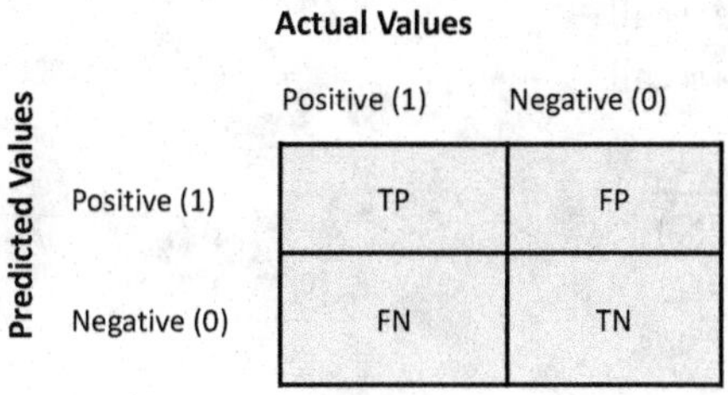

Actual Values

Figure 4.4 Confusion matrix.

Loss function

In supervised learning, *loss function* represents a measure of how far the predictions of a model are from the labels in the training data. Mean squared error (MSE) is a popular loss function.[8] MSE is computed by dividing the sum of squared differences by the number of datapoints used for training.

Confusion matrix

In binary classification, the true positive (TP), true negative (TN), false positive (FP), and false negative (FN) case counts are tabulated in a table called the confusion matrix (Figure 4.4). The matrix helps visualize the performance of the binary classifier.

Precision

Precision is defined as the frequency with which a model correctly predicts the positive class.

$$\text{Precision} = \frac{\text{TP}}{\text{TP+FP}} \tag{4.1}$$

Recall

The recall metric calculates how many of the actual positives the classifier can predict.

$$\text{Precision} = \frac{\text{TP}}{\text{TP+FN}} \tag{4.2}$$

The oft-used statistical F-measure is the harmonic mean of precision and recall, given by:

$$F = \frac{2}{\text{precision}^{-1} + \text{recall}^{-1}} \tag{4.3}$$

$$= 2\frac{\text{precision}*\text{recall}}{\text{precision}+\text{recall}} \tag{4.4}$$

$$= \frac{\text{TP}}{\text{TP}+\frac{1}{2}(\text{FP}+\text{FN})} \tag{4.5}$$

Overfitting

Overfitting or overtraining phenomenon is observed when the training accuracy is very high, and the subsequent testing accuracy is relatively low. Overfitting is the result of closely matching the model to the training dataset. It picks up the noise and peculiarities of the training dataset to such an extent that it fails to generalize the extracted data patterns. This results in poor performance on the test dataset.

Underfitting

Underfitting refers to the poor performance of a model on the training as well as test datasets. This happens when the model fails to capture the complexity of the training data. Training with too few data samples for too few epochs or at a low learning rate along with deficiencies in the model itself are some of the leading causes of underfitting.

DATA: THE FUEL FOR ML

Data is the driving force or the fuel for ML. Inexpensive and available in massive quantities, data has suddenly taken center stage in the enterprise world. Data serves as raw material for discovering patterns which lead to predictions and decision-making. Data analysts enable BO. Data cannot be used in its raw form for ML. The following section describes the data preprocessing essential for cleaning the raw data and transforming it into a format necessary for ML.

Data preprocessing

Field data is collected directly from the IoT sensors or put together from several web pages of the enterprise site or downloaded from a public repository. Web data is usually in multimedia format: text, numbers, speech, music, images, and video. The first step would be to separate the data into each respective medium. Text and/or number data is usually represented in a form of a table or spreadsheet containing horizontal rows and vertical columns. Each row, known in the data parlance as an *instance*, represents one example from the dataset. Each column, known as a feature or attribute, represents the observed or measured properties of the dataset. Finally, the intersection of any row and column is a cell which holds the value of the attribute or feature variable in that column.

The raw data may be in the form of figures (numbers) or labels (characters or words). The column headings indicate the features or the attributes of the data. They are represented as variables. In general, variables in data science could be qualitative or quantitative. Qualitative or categorical data like gender, nationality, and blood group is in the form of letters or labels, while quantitative or numerical data like age, weight, and income is in the form of numbers (Figure 4.5). Categorical data are nominal (unordered labels like nationality, gender), ordinal (ordered labels like grade, ranking), and binary (only two values like true/false, yes/no). While processing, it is customary to convert the Yes/No or True/False values of the binary variables into 0/1. Numerical values are discrete (set of integers) or continuous (set of real numbers).

Data cleaning

Raw data is not clean. It needs fixing because it may be messy, incomplete, and complex. Data cleaning is often the most tedious and most time-consuming step in the ML pipeline.

> Usually, data has to be moved, compressed, cleaned, chopped, sliced, diced, and subjected to any number of other transformations before it is ready to be used in the algorithms or visualizations that we think of as the heart of data science.[9]

Messy data

It contains noise in the form of wrong or corrupted values, out-of-range values, biases, contradicting values, typographical errors, and so on. Noise from data-gathering sensors, human errors in recording, and scanning from

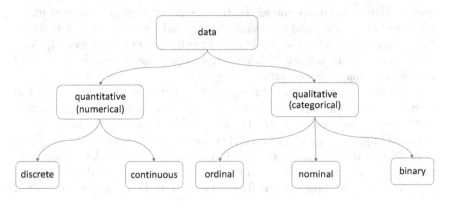

Figure 4.5 Qualitative and quantitative data.

old printed materials are some of the sources that bring in noise and corrupt the data. In the California housing dataset (Table 4.1), "ISLAND" for the geographical location of the house in place of "INLAND" is a typographical error that distorts the value of an attribute in a data record. Algorithms either break down when processing such faulty records, or produce non-sensical garbage results, which are often difficult to detect.

Incomplete data

Some rows (instances) and columns (features) may be missing in the data. Some values are just missed out by the sensors depending on the environmental conditions or other situational factors or by human error. The missing values are either imputed through statistical methods or the entire rows or columns with missing data are eliminated in the data cleansing process.

Complex data

There exist complex nonlinear relationships hidden in the data. These need to be sorted out through statistical methods. Sometimes, the sheer size and dimensions of the data can be a challenge for ML algorithms. Dimension reduction methods help simplify and project the data onto lesser dimensions without losing their internal relationships.

Feature selection

Features in a dataset are represented by columns when the data is in a tabular format as in the case of a spreadsheet. A feature is a numeric representation of an aspect of raw data. Features sit between data and models in the machine learning pipeline.[10] Feature selection is the process of reducing the number of features in a dataset. This results directly in the reduction in the number of input variables to an ML model. Feature selection relies on statistical methods for filtering noisy and irrelevant features and selecting only those which have a significant impact on analysis and prediction, or in other words, ML. The advantages of feature selection are computational cost reduction and model performance improvement.

There are two main types of feature selection techniques: supervised and unsupervised.[11] The major differences between the two approaches are that in the supervised approach, the target variable is used, while in the unsupervised approach the target variable is not used to perform feature selection. The unsupervised approach relies on statistical correlation methods to remove redundant variables in the dataset. The supervised methods are more elaborate and include the following three techniques: wrapper, filter, and evolutionary algorithms.

Wrapper

The wrapper methods systematically create subsets of features from the original dataset and record the effect of each subset on the performance of the learning model on a hold-out set. After repeated trials, a wrapper method retains the subsets which lead to relatively higher model performance. The main disadvantage of the wrapper method is that it tends to be exhaustive, and therefore, computationally expensive and time-consuming when dealing with datasets with a large number of features.

Filter

Filter methods, on the other hand, use statistical measures to select feature subsets. The measures chosen are fast to compute, while still capturing the relevance of the selected feature subset. Mutual information, chi-square, and Pearson correlation are some of the common filter measures.

Evolutionary algorithms

Evolutionary algorithms are a branch of algorithms in the computational intelligence (CI) paradigm.[12] Free from rigorous and rigid mathematical formulations, they rely on flexible computational constructs which derive their inspiration from natural phenomena. The genetic algorithm (GA), which has set the de facto standard for optimization algorithms, is a computational metaphor of Darwinian evolution. The algorithm computationally mimics the natural selection process[13] and solves even NP-hard optimization problems in a reasonable time window. The evolutionary algorithms begin with a population of randomly generated feasible solutions and iterating through fitness computation, selection, crossover, and mutation operators evolve the initial solutions into optimal ones. When an evolutionary algorithm is applied to feature selection, it selects a subset of features which optimizes the prediction accuracy.[14]

SUPERVISED LEARNING

Supervised learning acknowledges a supervisor that guides the learning agent. The learning agent is the ML algorithm or model, and the output in the data acts as a supervisor for a given set of inputs. The aim of the learning algorithm is to predict how a given set of inputs leads to the output. At first, the ML agent takes the inputs and randomly predicts the corresponding outputs. Since the random calculation is akin to shooting in the dark, the predicted outcomes stray far away from the known outcomes. The supervisor at the output end indicates the error in prediction which again guides the learning agent to minimize the error.

Learning proceeds through a large number of cycles called epochs and stops when the total error is minimized to almost zero. This phase of learning is referred to as training. The learned model obtained at the end of the learning phase is then tested using new and fresh data which may not contain any output for a given input record. The goal of the learned model is to predict the output given the input within a small margin of error. Without loss of generality, the more the data in training, the greater is the prediction accuracy of the learned model.

The following subsections describe two different supervised learning methods, one for continuous data and the other for discrete data.

Linear regression

Regression is a classical statistical method for determining relationship between the set of dependent variables and the independent variables which take continuous (real number) values. Once the functional form is created through learning from data, one can just plug in the input variables in the equation to compute the predicted value of the output variable. Regression is widely used in finance and investment, for example, to predict sales based on diverse factors such as GDP growth, previous sales, weather conditions, and so on. "Regression analysis helps businesses to answer questions such as: Which factors matter most? Which can be ignored? How do those factors interact with each other? And, perhaps most importantly, what is the level of certainty in those interacting factors?"[15]

Since regression is all about the functional relationship among the variables in the data, it is essential to have a thorough understanding of the concept of variables before undertaking any regression study. Refer to the explanation of dependent and independent variables in Box 4.1.

There are over 15 types of regression methods used in statistics and data science. In this chapter, the basic regression model frequently used in data science and business analytics, namely, linear regression, is considered. The following subsections explain simple linear regression and multiple linear regression.

Simple linear regression

This method is used when there is only one independent variable which influences the dependent variable. Given a dataset, if some kind of a rising or falling pattern is observed in unison in the values of the two variables, even without an expert domain knowledge it may be assumed that there exists a kind of linear relationship between the two variables.

A simple 2D scatter graph is plotted to verify the relationship between the two variables in the dataset. It is customary in statistical analysis to represent the dependent variable on the horizontal axis (x-axis) and the independent variable on the vertical axis (y-axis). The horizontal axis serves as the

BOX 4.1 VARIABLES AND THEIR TYPES

Variable represents some measure or quantity that can take varying values, one at a time. In mathematics and statistics, it is common to use the letters x, y, z, and so on to indicate variables. For example, age, weight, and height are simple variables in personal data; prices, indexes, and interest are variables in finance.

Dependent variable – This is the principal quantity or factor of interest to predict using the regression method. The dependent variable is usually denoted by y.

Independent variables – These are the quantities or factors (normally, more than one) that are assumed to have an impact on the dependent variable. The independent variables are commonly denoted as x_i.

baseline on which one can slide the value of the independent variable left or right to see the corresponding value on the vertical axis.

Figure 4.6 shows the scatter of points when the two variables are plotted on their respective axis. From the plot, it can be visually verified that when there is an increase in the independent variable, there is a corresponding increase in the dependent variable. In other words, there is a correlation

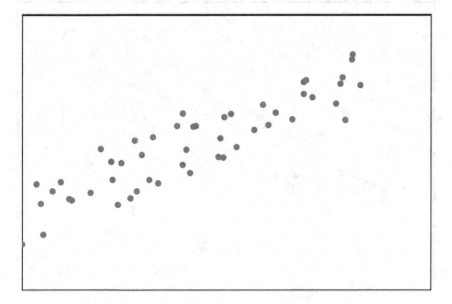

Figure 4.6 Plot of dependent variable vs independent variable.

between the values of the two variables. Linear regression aims to find the exact relationship by drawing a line in between the points such that the mean deviation of the points from the line is minimal. In statistical terms, linear regression is all about fitting a straight line to data.

The general functional form is assumed to be a line as given below:

$$y = a + bx + \varepsilon \qquad (4.6)$$

where:
  y=the variable to predict (dependent variable)
  x=the variable being used to predict y (independent variable)
  a=the intercept
  b=the slope
  ε=the regression residual (error)

There are detailed methods of determining the coefficients a, b, and ε from the dataset. Refer to Appendix A for software frameworks that readily compute the values of these coefficients with a mere input of the two columns of the dataset.

Figure 4.7 shows the regression line obtained by determining the regression coefficients from the data distribution. The line represents how much y changes with respect to any given change of x. The slope of the line given by the b parameter indicates whether the correlation is positive or negative.

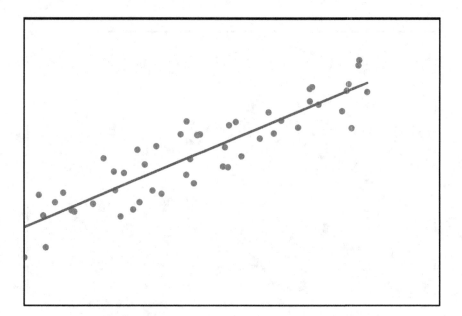

Figure 4.7 Linear regression line.

For a positive correlation, the line slopes upward from left to right, and for a negative correlation, it slopes downward from left to right.

It is straightforward to use the regression line to make predictions: plug in the desired value of x in the regression equation and compute the corresponding value of y; alternatively, draw a vertical segment from the desired coordinate x to the line and from there a horizontal segment to the vertical axis for visual verification of the predicted value.

Multiple regression

The real-world business situations are so complex that simple regression cannot cover them all. In such cases, the theory and analysis developed in the previous section are extended. Multiple regression is the extension of simple linear regression to multivariate data.

Consider two predictor variables and the predicted variable. A scatter can be plotted using a set of these three variables: x, y, z. The 2D scenario of simple regression described in the previous section progresses onto a 3D scenario. Here, a plane is drawn separating the data points (Figure 4.8). As the dimensions of the problem (number of predictor variables) increase, the analysis is extended to hyperplanes of higher dimensions. Although they cannot be visualized in practice, they can be accurately described through mathematical equations.

The regression idea can be further extended to include n number of independent or predictor variables. The equation for such a multiple-regression is:

$$y = a + b_1x_1 + b_2x_2 + \cdots\cdots b_nx_n + \varepsilon \qquad (4.7)$$

A real-world example for learning multiple regression is the California housing prices dataset, a small part of which shown in Table 4.1. The entire dataset is available in the public domain.[16]

In all, there are ten input features that determine the price of a housing unit. The features take on real values with the exception of the locality which takes nominal values, as explained below:

The features are as follows:

1. Longitude – A measure of how far west a house is; a higher value is farther west.
2. Latitude – A measure of how far north a house is; a higher value is farther north.
3. Housing median age – Median age of a house within a block; a lower number is a newer building.
4. Total rooms – Total number of rooms within a block.
5. Total bedrooms – Total number of bedrooms within a block.
6. Population – Total number of people residing within a block.

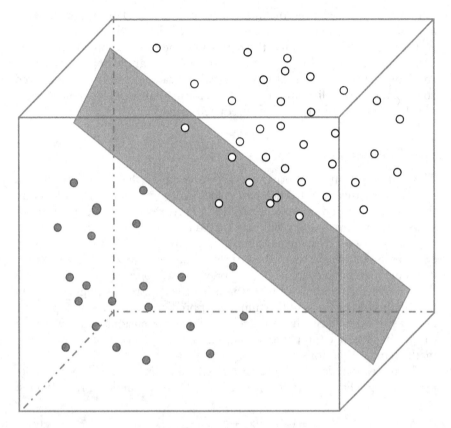

Figure 4.8 Linear regression with two independent variables.

    7. Households –Total number of households, a group of people residing within a home unit, for a block.
    8. Median income – Median income for households within a block of houses (measured in tens of thousands of US dollars).
    9. Ocean proximity – Location of the house with respect to the ocean/sea.
   10. Median house value – Median house value for households within a block (measured in US dollars).

The median house value shown in column 10 depends on the values of the 9 different feature values. The dataset needs some preprocessing and cleaning before passing it through ML algorithms to learn prediction. Some entries are missing. In such a case, the data can be imputed or the respective instance (row) can be eliminated. The entire column where the values are missing can also be deleted resulting in filtered data. For example, in row 6, the value of *total bedrooms* is missing. Since this record belongs to the *Luxury* housing class, we insert the average value of *total bedrooms* in

Table 4.1 California housing dataset

Rows	Longitude	Latitude	Housing median age	Total rooms	Total bedrooms	Population	Households	Median income	Ocean proximity	Median house value
1	−124.17	40.8	52	1,557	344	758	319	1.8529	NEAR OCEAN	62,500
2	−117.05	33.01	17	3,430	425	1468	433	10.6186	<1H OCEAN	429,300
3	−117.12	32.71	24	421	101	396	113	0.6433	NEAR OCEAN	111,300
4	−122.06	37.67	22	3,882	816	1830	743	4.2733	NEAR BAY	180,700
5	−119.67	36.81	4	1,262	216	622	199	4.9432	INLAND	114,400
6	−121.75	37.11	18	3,167	513	1414	482	6.8773	<1H OCEAN	467,700
7	−122.25	37.8	43	2,344	647	1710	644	1.6504	NEAR BAY	151,800
8	−117.87	33.75	18	697	255	812	221	2.6635	<1H OCEAN	162,500
9	−118.28	33.95	40	2,044	538	2150	524	2.1437	<1H OCEAN	94,800
10	−118.48	33.43	29	716	214	422	173	2.6042	ISLAND	287,500
11	−119.34	34.39	27	669	131	314	106	2.4659	NEAR OCEAN	231,300
12	−122.44	37.77	52	2,002	520	939	501	3.2239	NEAR BAY	488,900
13	−122.26	38.31	33	4,518	704	1776	669	5.2444	NEAR BAY	281,100
14	−116.87	33.76	5	4,116	761	1714	717	2.5612	INLAND	130,800
15	−122.33	38.39	36	831	122	272	109	6.3427	INLAND	304,500
16	−118.61	34.38	2	5,989	883	1787	613	6.6916	INLAND	329,500
17	−115.73	33.35	23	1,586	448	338	182	1.2132	INLAND	30,000
18	−119.22	34.15	32	3,152	596	3490	526	2.725	NEAR OCEAN	450,000
19	−118.28	34.16	49	1,393	290	605	282	2.9491	<1H OCEAN	257,400
20	−118.83	34.14	16	1,316	194	450	173	10.1597	NEAR OCEAN	500,001

Table 4.2 California housing dataset preprocessing of nominal labels

Near bay	Near ocean	<1h ocean	Inland
1	2	3	4

the luxury housing class in the missing slot. In row 10, the ocean proximity value is recorded as "*ISLAND*," which is obviously a misspelling of "*INLAND*."

The ocean proximity feature is a nominal feature. To aid the calculations, the nominal labels are converted into relative scores as shown in Table 4.2.

Now that clean data is available, ML experiments can be conducted to perform classification. The multiple regression equation for the above problem is framed as:

$$y = a + b_1x_1 + b_2x_2 + \cdots\cdots b_9x_9 + \varepsilon \qquad (4.8)$$

Solving the above equation using the dataset will give the values of the regression coefficients and the other unknowns. Some of the ML computational frameworks mentioned in Appendix A readily yield a solution. The parametric function can then be used to predict the housing price for any new data record.

Neural networks

Neural network (NN) has proved to be the most successful supervised learning method. Deep learning which has gained spotlight in ML is almost entirely based on NN models. NN is modeled on the mammalian brain. The individual cells in the biological brain, called neurons, are interconnected to form a large neural network. Each neuron possesses an input and an output filament called a dendrite and an axon, respectively. The dendrites provide input signals to the cell, and the axon transmits output signals to other neurons to which it is connected. Signals can be transmitted unaltered or they can be altered by the synapse which is able to increase or decrease the strength of the connection among the neurons. It is the synapse that causes excitation or inhibition of the subsequent neurons.

NN are multilayered networks consisting of at least three different layers: the input layer, the hidden layer, and the output layer (Figure 4.9). The number of neurons in the input as well as in the output layer is determined by the problem the NN is learning to solve. However, the number of neurons in the hidden layers is arbitrary.

Each neuron in a layer is connected to every other neuron in the successive layer. The connection between any two neurons has an associated weight. In essence, the learning carried on by the network is encapsulated

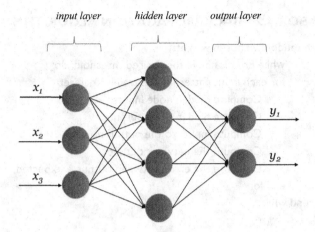

input layer   *hidden layer*   *output layer*

Figure 4.9 Artificial neural network.

in these interconnection weights. Each neuron calculates a weighted sum of the incoming neuron value, transforms it through an activation function, and passes it on as the input to subsequent neurons. The information processing proceeds from the input layer to the output layer via the hidden layer – hence the name, *Feedforward Network*.

The n number of inputs in the input layer of the ANN is represented as a vector:

$$X_i = (X_1, X_2 \dots X_n) \qquad (4.9)$$

The weights on the connection arcs can be represented by the following weight matrix:

$$W_{ij} = \begin{pmatrix} W_{11} & \cdots & W_{1m} \\ \vdots & \ddots & \vdots \\ W_{n1} & \cdots & W_{nm} \end{pmatrix} \qquad (4.10)$$

where n is the number of neurons in the ith layer and m is the number of neurons in the next j^{th} layer.

The summation of the inputs at the hidden layer neurons is:

$$S_j = \sum_{i=1}^{n} \sum_{j=1}^{m} X_i W_{ij} + b \qquad (4.11)$$

where b is the bias.

BOX 4.2   BACKPROPAGATION ALGORITHM

Start with randomly chosen weights;
      while *Error* is above the desired threshold, do:
      for each input pattern in the training dataset:
         Compute hidden node inputs;
         Compute hidden node outputs;
         Compute inputs to the output nodes;
         Compute the network outputs;
         Compute the error between output & expected;
      end for
    end while
  End

The sum of the weights are passed through an activation function leading to nonlinear outputs of the hidden layer neurons. In most cases, the sigmoid function is used as the activation function.

If t_i is the expected output of the ith output neuron and y_i is its actual output, then the total error at the output of the neural net is:

$$Error = \frac{1}{2}(t_i - y_i)^2 \qquad (4.12)$$

The backpropagation algorithm for training the neural net, given a training dataset, is stated in Box 4.2

Classifying California housing prices using NN

In real life, most house-buyers will be interested in knowing the exact price of a property. However, consider a variation of the problem which any real estate retailer may propose.

Instead of quoting the exact price of the property, which may fluctuate over a period of time, the output data is presented in the form of classes like low, lower middle, middle, upper middle, and luxury. The following table gives the housing classes along with their respective lower and upper bound of prices.

The transformed housing dataset will have nine input features and six classes as outputs. A simple NN is constructed to perform classification. Since there are nine inputs, the NN input layer will have nine neurons. As a rule of thumb, the number of neurons in the middle layer will be $2n+1 = 2 \times 9 + 1 = 19$. The experiment begins with just a single hidden layer

Table 4.3 California housing dataset classes

Lower bound of price	Higher bound of price	Housing category
14,999	100,000	Low (Lo)
100,000	160,000	Lower Medium (LM)
160,000	240,000	Medium (M)
240,000	310,000	Upper Medium (JIM)
310,000	450,000	Above Medium (AM)
450,000	500,001	Luxury (Lx)

to keep it simple. Later, the number of hidden layers may be varied and the number of neurons in each of the hidden layers, too. As dictated by the dataset, the NN outputs will be six, each output representing a unique class. In the modified dataset, the median price attribute is replaced by the house class (Table 4.4).

The dataset is randomly divided into a 70% training subset and a 30% test subset. The NN is then trained with the training set using the back-propagation algorithm shown in Box 4.2. The training is stopped when the learning model reaches an adequate level of convergence. The learned model is then used to make predictions using the test dataset.

UNSUPERVISED LEARNING

As seen in the previous section, supervised learning assumes the data is neatly labeled. For a vector of input elements, there is a corresponding output value. The learning algorithm learns a mapping that maps the inputs to the output. For every set of input values, the learning process is guided by the output value, which acts as a supervisor. But what will happen if a dataset contains only the inputs, without any corresponding outputs? In fact, most real-world data is usually unlabeled, like for instance, the data of customers or online shoppers. Labeling data is a time-consuming and costly endeavor. Besides, the activity needs domain expertise. Is there a way of performing ML on unlabeled data?

Unsupervised learning precisely addresses that issue. It discovers unknown patterns in the data without having any external guidance or supervision and arranges the data into clusters. A cluster is a group of datapoints that are similar and, at the same time, dissimilar to the datapoints in other clusters. In short, USL sorts out unsorted data based on the feature similarities and differences even though there are no categories provided.

For example, the dataset of objects shown in Figure 4.10a has a set of features for each object, but there is no corresponding label indicating what the object is. When an unsupervised learning algorithm is presented with the features of the dataset objects, it discovers the patterns in the data and

Table 4.4 California housing dataset modified for classification

Rows	Longitude	Latitude	Housing median age	Total rooms	Total bedrooms	Population	Households	Income	Ocean proximity	Class
1	-124.17	40.8	52	1557	344	758	319	1.8529	NEAR OCEAN	Low
2	-117.05	33.01	17	3430	425	1,468	433	10.6186	<1H OCEAN	AM
3	-117.12	32.71	24	421	101	396	113	0.6433	NEAR OCEAN	LM
4	-122.06	37.67	22	3882	816	1,830	743	4.2733	NEAR BAY	M
5	-119.67	36.81	4	1262	216	622	199	4.9432	INLAND	LM
6	-121.75	37.11	18	3167	513	1,414	482	6.8773	<1H OCEAN	Lux
7	-122.25	37.8	43	2344	647	1,710	644	1.6504	NEAR BAY	LM
8	-117.87	33.75	18	697	255	812	221	2.6635	<1H OCEAN	M
9	-118.28	33.95	40	2044	538	2,150	524	2.1437	<1H OCEAN	Low
10	-118.48	33.43	29	716	214	422	173	2.6042	ISLAND	UM
11	-119.34	34.39	27	669	131	314	106	2.4659	NEAR OCEAN	M
12	-122.44	37.77	52	2002	520	939	501	3.2239	NEAR BAY	Lux
13	-122.26	38.31	33	4518	704	1,776	669	5.2444	NEAR BAY	UM
14	-116.87	33.76	5	4116	761	1,714	717	2.5612	INLAND	LM
15	-122.33	38.39	36	831	122	272	109	6.3427	INLAND	UM
16	-118.61	34.38	2	5989	883	1,787	613	6.6916	INLAND	AM
17	-115.73	33.35	23	1586	448	338	182	1.2132	INLAND	Low
18	-119.22	34.15	32	3152	596	3,490	526	2.725	NEAR OCEAN	AM
19	-118.28	34.16	49	1393	290	605	282	2.9491	<1H OCEAN	UM
20	-118.83	34.14	16	1316	194	450	173	10.1597	NEAR OCEAN	Lux

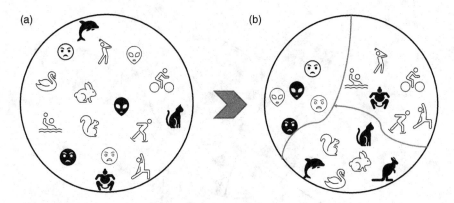

Figure 4.10 (a) Original dataset. (b) Clustered dataset.

divides the original dataset into three clusters as shown in 4.10b such that the objects belonging to a cluster are similar to one another and dissimilar to the objects in the other clusters.

Unsupervised learning has been in use in the data mining and statistical community for quite a while before the advent of the AI ML paradigm. Out of the numerous USL algorithms found in the literature, this section presents the k-means clustering and the DBSCAN algorithms.

k-means

k-means is one of the most popular unsupervised ML algorithms. The k-means algorithm arranges data into a fixed number (k) of clusters such that the datapoints in a given cluster are similar to one another and, at the same time, dissimilar to the datapoints in other clusters.

The k-means clustering is a popular unsupervised ML algorithm which has been in use in the data mining community for a long time. Since the output (label/supervisor) for each of the input record (feature set) in the dataset is not known, the algorithm must learn to cluster the datapoints based on the feature space. The algorithm iteratively partitions the datapoints into k clusters such that each datapoint belongs to the cluster with the nearest mean or cluster centroid. Effectively, it discovers patterns in the data and groups similar datapoints together. The parameter k from which the k-means clustering algorithm derives its name is fixed a priori.

The k-means algorithm begins with a first group of randomly selected centroids which serve as the starters for every cluster. The algorithm then goes through a cycle of predetermined number of iterations, performing calculations to optimize the positions of the centroids in each iteration. It halts either when the predetermined number of iterations have been reached or when the centroids have stabilized.

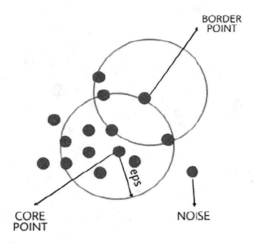

Figure 4.11 DBSCAN algorithm points.

Density-based spatial clustering of applications with noise (DBSCAN)

Unlike the k-means clustering, DBSCAN does not require prior knowledge of the number of clusters. The DBSCAN algorithm considers the two important parameters and three types of points (Figure 4.11)

Eps – serves as a threshold to determine the neighborhood around a datapoint. If the distance between two points is lower than or equal to eps, these points are considered as neighbors. The performance of the algorithm depends on the optimal value of *eps*. If it is too small, then many datapoints will be considered as outliers. If it is too large, then the clusters will merge and the majority of the datapoints will fall in the same clusters.

minPoints – refers to the minimum number of neighbors within the *eps* radius. In practice, the value of minPoints is set in proportion to the size of the dataset. As a rule, minPoints ≥ d+1, where d is the number of dimensions of the datapoints. The minimum value of minPoints must be 3.

In this algorithm, there are three types of data points.

Core point – A point which has more than minPoints points within the eps radius.
Border point – A point which has less than minPoints within the eps radius, but is in the neighborhood of a core point.

Noise – A point which is neither a core point nor a border point.

Density edge – If c1 and c2 are core points and d(c1,c2)≤eps, then the edge connecting c1 and c2 is called the density edge.

Density connected points – Two points a and b are said to be density connected points if both a and b are core points and there exists a path formed by density edges connecting point (a) to point(b).

The DBSCAN algorithm works as follows[17]:

Step 1 – Using the eps and minPoints criteria, label all datapoints as core, border, or noise points.

Step 2 – Remove all noise points as they do not belong to any clusters.

Step 3 – For every core point p that has not yet been assigned to a cluster:
   a.   Create a new cluster with the point p.
   b.   add all the points that are density-connected to p.

Points a and b are said to be density connected if there exists a point c which has a sufficient number of points in its neighbors and both the points a and b are within the eps distance.

Step 4 – Assign each border points to the cluster of the closest core point.

Step 5 – Iterate through the remaining unvisited points in the dataset. Those points that do not belong to any cluster are considered noise.

Semi-supervised learning

Semi-supervised machine learning is a combination of supervised and unsupervised machine learning methods. In supervised machine learning, the learning model is trained on a "labeled" dataset. This allows the model to identify relationships between the target variable and the rest of the variables in the dataset based on the information it has gathered during training. Unsupervised machine learning algorithms, on the other hand, learn to cluster a dataset without the outcome variable. In semi-supervised learning, an algorithm learns from a dataset that includes both labeled and unlabeled data.

Recall that SL is used when the dataset is labeled and USL is used when the data is not labeled. SSL is a combination of these two categories of ML. SSL becomes useful when there is a large dataset and only a small part of it is labeled. It is laborious, time-consuming, and extremely costly to get all the data labeled. When faced with a situation where there is a large dataset with only a few labeled instances, the semi-supervised learning is used. The two main techniques in SSL, namely, self-training and co-training are described below:

Self-training

Self-learning is known by alternative terms such a self-teaching, self-labeling, or decision-directed learning. Self-training is considered to be a single-view supervised algorithm because it relies on its own predictions on unlabeled data to teach itself.[18] The self-training method first trains the ML algorithm on the small dataset that is already labeled. This training is the same as one would do in a totally SL scenario. Once the model is fully trained on the small collection of labeled data, it is then used to predict the labels of the unlabeled data in the entire collection. The process of labeling the initially unlabeled data by training a model on the partially labeled data is sometimes called pseudo-labeling. There are variations on pseudo-labeling. The entire subset of unlabeled data could be labeled or only those with a higher confidence.

After completing the stage of labeling all the instances, the model is retrained on the entire dataset.

Self-training is a very simple method often used in Natural Language Processing. Text documents available on the Web are huge volumes of scripts, books, blogs, ads, and so on which are mostly unlabeled. Professional librarians and people in charge of downloading and classifying various forms of Web text documents cannot possibly cope with reading all the text and then manually classifying the documents. Self-training helps in classifying the large collection of unlabeled documents when only a handful of labeled documents are available.

SSL works on the implicit assumption that the labeled data and unlabeled data are independent and identically distributed. The biggest disadvantage of self-training is that it is unable to correct itself. It can lead to poor results because some of the earlier mistakes in classification can reinforce themselves.

Co-training

Co-training assumes that[19]:

i. features can be split into two sets independent of one another.
ii. each sub-feature set is sufficient to train a good classifier.

Initially, two separate classifiers are trained with the labeled data on the two sub-feature sets, respectively. Each classifier then classifies the unlabeled data and "teaches" the other classifier with the few unlabeled examples (and the predicted labels) they feel most confident. Each classifier is retrained with the additional training examples given by the other classifier, and the process repeats.

Co-training is based on the assumption that the two models are complementary to one another and can keep on correcting each other iteratively till they reach convergence.

In the case of web pages classification, for example, words on the webpage and the links that point to that page are used as two different sets of features.[20] Accordingly, co-training algorithms that learn to classify web pages use the text on the page as one feature set (or one view) and the anchor text of hyperlinks on other pages that point to that page as another feature set (or the other view). The two models teach one another through co-training and determine the likelihood that a page will contain data relevant to the search criteria.

Another well-established co-training technique is through using data augmentation.

Tri-training

Zhou and Li (2005)[21] propose an alternative multi-view training method. Three different classifier models are trained on three different data subsets obtained from the original labeled dataset through the well-established statistical method of bootstrap sampling. The three models m1, m2, and m3 are simultaneously trained on these bootstrap subsets. An unlabeled data point is added to the training set of a model if the other two models agree on its label. Training is carried on till the classifiers do not produce any significant change in pseudo-labeling.

REINFORCEMENT LEARNING

With the increased presence of AI in the gaming industry, developers are challenged to create highly responsive and adaptive games by integrating artificial intelligence into their projects.[22] Reinforcement learning (RL) techniques and algorithms play an important role in game development.

Loosely speaking, RL is very similar to teaching a dog to learn new tricks. When the dog performs the trick as directed, it is rewarded; when it makes mistakes, it is corrected. The dog soon learns the tricks by mastering the policy of maximizing its rewards at the end of the training session. Technically, RL algorithms are used to train biped robots to walk without hitting into obstacles, self-driving cars to drive by observing traffic rules and avoiding accidents, and software programs to play games against human champions and win. The agent (software program engaged in learning) is not given a set of instructions to deal with every kind of situation it may encounter as it interacts with its environment. Instead, it is made to learn the art of responding correctly to the changing environment at every instant of time.

RL consists of a series of states and actions. The agent is placed in an environment which is in some state at some time. When the agent performs an action, it changes the state of the environment which evaluates the action and immediately rewards or punishes the agent[23] (Figure 4.12).

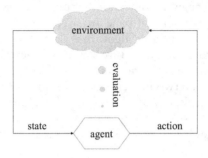

Figure 4.12 RL framework.

The following are some of the key elements in reinforcement learning:

Agent – Computer program engaged in reinforcement learning. The salient feature of an agent is that it is gifted with autonomy, which means it can make its own decisions without being explicitly programmed. The agent learns a policy for taking actions at every time step after observing the environment.

Environment – Representation of the world in which the agent operates. The agent can perceive the environment and act on it, but it has no control over it. In a board game, the board configuration is the environment; in autonomous driving, the road, traffic, and immediate physical surroundings form the environment; in finance, the market becomes the environment. It is the environment that gives reward to the agent on evaluating its actions.

State – The current situation of the agent. The agent uses the state information to decide (autonomy) upon the next action. For example, the state could contain variables that characterize the market, such as inventory of shares of stock, market quality measures, and price volatility.

Action – A single move or change initiated by agent which changes its state and the environment. In a board game, any move made by the agent/player, in autonomous driving, move/stop/ turn, and in finance, a decision to invest are some of the relevant actions taken by the agent.

Policy – A mapping from the states of the environment to the probabilities of selecting actions that can be taken in those states. It represents the behavior function of the agent.

Reward – Score received by the agent in response to its actions. For example, in finance applications, the rewards could reflect the change in profit made by taking a particular action.

Value function – The sum of expected reward an agent will receive in the future starting from a particular state. The estimation of the value function is important for estimating the greatest reward to be received in the long run.

A typical RL algorithm tries mapping observed states onto actions, while optimizing a short-term or a long-term reward. Interacting with the environment continuously, the RL agent tries to learn policies, which are sequences of actions that maximize the total reward. The policy learned is usually an optimal balance between the immediate and the delayed rewards.

The overall reward function consists of immediate rewards and delayed rewards. The immediate reward reflects the impact of the current action, while the delayed reward reflects the impact of the action on future states of the environment. The delayed reward is accounted for using the discount factor γ which takes a value between zero and one. Lower values of γ make the agent shortsighted embracing the immediate rewards, while higher values tip the system toward long-term rewards.

The two major categories of RL are model-free RL and model-based RL (Figure 4.13). As seen in the preceding sections, the learning agent in a given state acts on the environment and receives a reward. The biggest challenge in RL is to predict the changes in the environment over which the agent has no control whatsoever. Learning a policy function by acting over the changing environment in a series of steps takes a phenomenal amount of time. This is the approach taken by model-free RL agents, who do not have any model of the environment to base their action decisions on. Most autonomous driving RL programs are model-free. In game playing, on the other hand, the enclosing boundaries of the playing field, board configurations, and the rules of the game provide a specific model of the environment in which the agent learns to act. Rather than acting on a raw and unknown environment, the model-based agents are embedded with a model of the environment. They sample their decisions and actions and learn to maximize the value function. The model-based RL is speedier than the model-free RL because it decreases the number of interactions with the environment by building a model of the environment and using it during training.[24] Model-free RL is further divided into Q-Learning and Policy Optimization, while model-based RL is further divided into "Given the model, learn by acting" and "Learn the

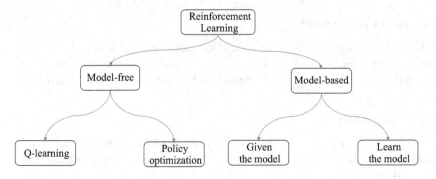

Figure 4.13 RL algorithm types.

model by acting." Since Q-learning is frequently used as an RL algorithm, it is described below.

Q learning

In Q learning, the agent tries to maximize the cumulative reward at the end of a learning cycle. The Q value (i.e., *Quality Value*) is given by the following equation:

$$Q(s,a) = r(s,a) + \gamma \; \max_{a} \; Q(s',a)$$

The above equation, known as the Bellman equation, computes the overall reward for the actions performed by the learning agent on the environment. It consists of the immediate reward for the action *a* corresponding to the state *s* and the maximum reward for choosing the appropriate future action in the future state *s'*. The discount factor γ, as explained above, is introduced in the equation to act as a balancing factor between the immediate and long-term rewards.

The Q-learning algorithm proceeds by keeping a Q-table which contains the rewards for any action the agent may take given the state. The Q-table entries are updated using the Bellman equation.

Financial applications of RL

RL has given the impression that it has impacted nothing more than the gaming world. It is true that the major impact of RL has been in developing computer games and in designing powerful algorithms to beat human world champions in board games; nevertheless, RL has shown significant success in finance, too. The following are three areas where RL models are employed successfully.

Portfolio optimization

RL algorithm aims to optimize asset allocation for portfolios to help investors achieve their financial goals. Typically, expected returns are maximized while risks are minimized taking into consideration the financial constraints.

Optimal trading

Q-network algorithms are often used to design an online trading system that continually monitors the stocks and learns to generate profit from trading in financial markets.

Recommendation systems

Traditionally, collaborative filtering and content-based methods are used to recommend items to online shopping. However, these systems are not very efficient. The recent RL recommender systems learn from customer data automatically collected online and constantly provide shopping suggestions and options to online customers. They are efficient and overcome the cold-start problem plaguing the traditional recommender systems. Moreover, their continual online learning of customer behavior results in explainable recommendations which further improve customer satisfaction.

CONSOLIDATION WORKSHOP

1. How does ML differ from traditional computing? Explain with an example.
2. How is ML fundamentally different from human learning? Discuss with an example.
3. Discuss, with an example, the six steps of ML implementation (*Hint:* Figure 4.1).
4. What is the difference between parametric and nonparametric models?
5. What is training, validation, and testing of ML models? What are the likely business issues in each of these aspects?
6. Describe the differences between qualitative and quantitative data.
7. Describe the detailed steps in data cleaning and preprocessing for ML.
8. What is the key difference between supervised and unsupervised learning? How can each type benefit businesses?
9. Describe the attributes of neural networks. How close are they to biological neural networks? How do they lead to ML models?
10. Explain the k-means algorithm with examples.
11. What is the DBSCAH algorithm?
12. What are the characteristics of reinforcement learning? What is the role of rewards and penalties in RL?
13. What is "Q" in reinforcement learning and what is its importance from a business perspective?
14. Distinguish between model-free and model-based RL algorithms.
15. What is the importance of the discount factor in Q learning?

NOTES

1. http://ww.vordenker.de/downloads/turing-vorlesung.pdf.
2. What is machine learning? The Conversation, (2017, May 4), https://theconversation.com/what-is-machine-learning-76759.
3. Michel, T. *Machine Learning*, McGraw-Hill, New York, NY (1997 October 1).

4. Loh, Wei-Yin. "Classification and regression trees." *Wiley Interdisciplinary Reviews---Data Mining and Knowledge Discovery* 1(1) (2011): 14–23.
5. Pedregosa, F. (2016). Hyperparameter optimization with approximate gradient. *arXiv preprint arXiv:1602.02355.*
6. Russell, S. and Norvig, P. *Artificial Intelligence: A Modern Approach*; 3rd ed., (2009).
7. Kuhn, Max and Johnson, Kjell. *Applied Predictive Modeling*, Springer; 1st ed., (2013), Corr. 2nd printing 2018 Edition, New York, NY (2013, May 17).
8. James, Gareth, Witten, Daniela, Hastie, Trevor, and Tibshirani, Robert. *An Introduction to Statistical Learning---With Applications* in R, Springer, New York, NY (2013).
9. Squire, Megan. *Clean Data*, Packt Publishing, Birmingham, UK (2015).
10. Brownlee, Jason. *Feature Engineering for Machine Learning*, (2018).
11. Brownlee, Jason. *Data Preparation for Machine Learning*, (2020).
12. Eberhart, Russell C. and Shi, Yuhui. *Computational Intelligence: Concepts to Implementations*, Morgan Kaufmann Burlington, MA; 1st ed., (2011, April 18).
13. Eiben, A.E. and Smith, J.E. *Introduction to Evolutionary Computing* (Natural Computing Series), Springer, New York, NY, 2nd ed., (2015, July 2).
14. Gonsalves, Tad. Feature subset optimization through the fireworks algorithm. *International Journal of Electronics and Computer Science Engineering.* ISSN-2277-1956.
15. https://hbr.org/2015/11/a-refresher-on-regression-analysis).
16. https://www.kaggle.com/camnugent/california-housing-prices.
17. DBSCAN Clustering Algorithm in Machine Learning (20:n17) https://www.kdnuggets.com/2020/04/dbscan-clustering-algorithm-machine-learning.html.
18. Wu, D., Shang, M., Luo, X., Xu, J., Yan, H., Deng, W., and Wang, G. (2018). Self-training semi-supervised classification based on density peaks of data. *Neurocomputing*, 275: 180–191.
19. Blum, A. and Mitchell, T. (1998, July). Combining labeled and unlabeled data with co-training. In *Proceedings of the Eleventh Annual Conference on Computational Learning Theory* (pp. 92–100).
20. Blum, A. and Mitchell, T. (1998, July). *Ibid.*
21. Zhou, Z. H. and Li, M. (2005). Tri-training: Exploiting unlabeled data using three classifiers. *IEEE Transactions on Knowledge and Data Engineering*, 17(11): 1529–1541.
22. Lanham, Michael. *Hands-On Reinforcement Learning for Games: Implementing Self-Learning Agents in Games Using Artificial Intelligence Techniques.* Packt Publishing, (2020, January 3).
23. Sutton, Richard S. and Barto, Andrew G. *Reinforcement Learning: An Introduction* (Adaptive Computation and Machine Learning series), A Bradford Book; 2nd ed., (2018, November 13).
24. Lapan, Maxim. Deep Reinforcement Learning Hands-On---Apply Modern RL Methods, with Deep Q-Networks, Value Iteration, Policy Gradients, TRPO, AlphaGo Zero and more, Packt Publishing, Birmingham, UK (2018, June 21).

Chapter 5

Dynamicity in learning
Smart selection of learning techniques

DYNAMICITY IN ML

Automated machine learning (ML) represents a fundamental shift in the way organizations approach data science. Current ML theory and practice has a challenge to handle. ML models are selected keeping the data and its format in mind; furthermore, ML models are also based on the business context. The context as well as the incoming data can change – leading to a potentially scattered ML model that may have lost its relevance. This chapter outlines an integrated online dynamic ML framework that unifies the scattered ML technology in data science.

Conventional ML algorithms are static. They feed on datasets that were handcrafted by experts prior to the ML training sessions. Many firms get accustomed to train their models on such old historical data. In reality, however, datasets keep changing at a rapid pace. Businesses need to align their models to cope with the online changing data. The dynamic ML framework takes ML to the next level. The framework proposed in this chapter responds to the changing data challenge semiautomatically. Through an expert system engine at the front end interface, it queries the user about the characteristics of the incoming data. The framework then inferences and chooses the appropriate ML category (supervised, unsupervised, etc.) and ML mode (shallow, deep, transfer learning) to match with the incoming data. Once set into motion online, the framework constantly monitors the changing data at the input, selects the appropriate category and mode of learning, and updates prediction at the output, without relying on the expertise of a professional data scientist.

Applying traditional machine learning methods to real-world business problems is a time-consuming, expensive, and error-prone endeavor. Such attempts require hiring a team of professional data scientists. These sought-after professionals are expensive and in short supply. Several data science reports lament that it is hard to hire a data scientist and harder still to keep her in the enterprise.[1,2]

Non-data scientists cannot build, train, test, and maintain the ML system. The size and complexity of systems is a daunting task. Expertise in programming and deep knowledge of mathematics are required. Managers and stakeholders who are not experts in data and ML can also use the dynamic ML framework to enhance and optimize their business processes.

A description and comparison of static and dynamic ML are provided next.

Static learning

Static learning is performed manually most of the time. Static learning is also known as offline learning. The datasets are almost always prepared offline. Relevant data is collected from various sources and checked by domain experts and data scientists for inconsistencies. The datasets are later preprocessed by using some of the techniques described in Chapter 4. Upon completion of the data preprocessing stage, the data is then divided into training, validation, and testing subsets. The data scientists then look for an appropriate model to be trained for prediction. The model is trained, validated, and tested using the predetermined subsets. The model hyperparameters are also tuned to give the desired level of prediction accuracy. The learned model is then frozen as a time-tested artifact. In static or offline learning, the model is trained only once and then used for prediction for a while.[3] When new enterprise data becomes available, the model is used for making predictions.

Static learning is easier and cost-effective as far as implementation and maintenance are concerned. The ML team undertakes a one-time effort and investment to obtain the trained model. The benefits can then be reaped against new data that steadily becomes available in the enterprise. On the downside, models that are trained in a static manner cannot cope with rapidly changing data. There are weekly, monthly, seasonal, and annual data changes in any enterprise. Consider, for instance, firms selling costumes and apparel. Their static machine learning models will not be in a position to predict sales due to seasonal variation in data. The statically learned models will certainly rupture around the time of Halloween, because of a huge variation in the sale of costumes. To respond to the data-driven business requirements, a dynamic learning framework is presented in the following sections.

Dynamic learning

Data in real life is not static. It is constantly changing, growing, and diversifying. Dynamic learning is designed to adapt to changing data. Older datasets are updated and newer datasets are created with the passage of time. Dynamic learning models are flexible to take the new forms of data and continue the learning task. This kind of online learning is adaptive and continuous. The ML algorithm continuously improves its learning and prediction is performed on the fly. The major disadvantage of the dynamic learning systems is that the incoming data, the model, and the entire ML pipeline have to be continually monitored.

DATA AND ALGORITHM SELECTIONS

Data is the oil for the machinery of ML. Everything in ML is data-driven. The type of learning and the ML algorithm to be chosen for implementing ML are all dictated by data. Data scientists and ML experts look at the data and manually select the appropriate ML algorithm. The criteria for selecting ML algorithm are input–output pairs, absence of output variable, few input–output pairs, absence of state-action-reward tuples, and data collection by interacting with the environment, as summarized below.

Input–output pairs

Datasets are usually in the form of tables with rows and columns. The columns represent the data attributes. The columns on the left represent the independent variables. The rightmost column is the output vector and represents the dependent variable. Supervised learning algorithms seek the functional relationship between the input (independent) and output (dependent) variables. Table 5.1 is an example of a dataset that can be used for regression as well as classification. It is related to the red variant of the Portuguese *Vinho Verde* wine.[4]

The rightmost column indicates the quality of red wine on a scale from 0 to 10, dependent on the 11 attributes or features given in the columns on the left. The dataset is also available in the UCI public repository.[5]

Absence of output variable

The data table is almost exactly as in Table 5.1, but without the rightmost column containing the output. The unsupervised learning algorithms seek patterns in the input feature space and group the data into clusters such that each cluster contains similar datapoints.

Few input–output pairs

Most real-life datasets do not have an output label with respect to the input. Labeling the data is expensive and time-consuming. However, with expert knowledge, some datapoints in the dataset are labeled. Semi-supervised algorithms learn classification from the labeled datapoints and assign pseudo-labels to the unlabeled datapoints. The pseudo-labels are refined iteratively through self-training or co-training (Table 5.2).

Table 5.1 Dataset format

#	Fixed acidity	Volatile acidity	Citric acid	Residual sugar	Chlorides	Free sulfur dioxide	Total sulfur dioxide	Density	pH	Sulphates	Alcohol	Quality
1	7.4	0.7	0	1.9	0.076	11	34	0.9978	3.51	0.56	9.4	5
2	7.8	0.88	0	2.6	0.098	25	67	0.9968	3.2	0.68	9.8	5
3	7.8	0.76	0.04	2.3	0.092	15	54	0.997	3.26	0.65	9.8	5
4	11.2	0.28	0.56	1.9	0.075	17	60	0.998	3.16	0.58	9.8	6
5	7.4	0.7	0	1.9	0.076	11	34	0.9978	3.51	0.56	9.4	5
6	7.4	0.66	0	1.8	0.075	13	40	0.9978	3.51	0.56	9.4	5
7	7.9	0.6	0.06	1.6	0.069	15	59	0.9964	3.3	0.46	9.4	5
8	7.3	0.65	0	1.2	0.065	15	21	0.9946	3.39	0.47	10	7
9	7.8	0.58	0.02	2	0.073	9	18	0.9968	3.36	0.57	9.5	7
10	7.5	0.5	0.36	6.1	0.071	17	102	0.9978	3.35	0.8	10.5	5
11	6.7	0.58	0.08	1.8	0.097	15	65	0.9959	3.28	0.54	9.2	5
12	7.5	0.5	0.36	6.1	0.071	17	102	0.9978	3.35	0.8	10.5	5
13	5.6	0.615	0	1.6	0.089	16	59	0.9943	3.58	0.52	9.9	5
14	7.8	0.61	0.29	1.6	0.114	9	29	0.9974	3.26	1.56	9.1	5
15	8.9	0.62	0.18	3.8	0.176	52	145	0.9986	3.16	0.88	9.2	5

Table 5.2 Criteria for selecting ML algorithms

Data type	Learning category
Input–output pairs	Supervised
Absence of output variable	Unsupervised
Few input–output pairs	Semi-supervised learning
Absence of state-action-reward tuples	Reinforcement learning
Data collection by interacting with environment	Deep reinforcement learning

Absence of state-action-reward tuples

In reinforcement learning, no data is provided to the learning agent. The agent perceives its current state and randomly takes an action. The action is immediately rewarded or penalized by the environment. Further, the agent's action changes the state of the environment and the agent takes the next action on the new stage. The state-action-reward cycles last till the end of a reinforcement learning session, where the cumulative reward is computed. In the absence of the state-action-reward data tuples, the agent must learn a policy that maximizes the long-term cumulative rewards.

Data collection by interacting with environment

Deep reinforcement learning methods, however, require active online data collection, where the model actively interacts with its environment. This makes such methods hard to scale to complex real-world problems, where active data collection means that large datasets must be collected for every experiment – this can be expensive and, for systems such as autonomous vehicles or robots, potentially unsafe.

In many domains of practical interest, such as autonomous driving, robotics, and games, there is an active online data collection when the model actively interacts with its environment. The previously collected interaction data consists of informative behaviors – state-action-rewards tuples. Deep RL algorithms that can utilize such prior datasets will not only scale to real-world problems, but will also lead to solutions that generalize substantially better.[6]

GAME TREE AND STATE EXPLOSION

The Min-Max algorithm is a robust artificial intelligence (AI) algorithm for developing game-playing software programs. Min and Max are metaphorical agents playing a board game. Max (computer program) is trying to maximize her score at every move of the game, while Min (human player) is trying to oppose and minimize Max's score. The algorithm draws a game

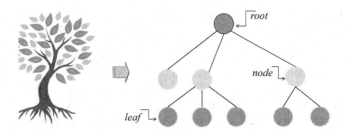

Figure 5.1 Tree structure.

Table 5.3 Game tree complexity of various board games

Game	State-space complexity	Game-tree complexity	Branching factor	Average game length
Tic-Tac-Toe	10^3	10^5	4	9
Connect four	10^{13}	10^{21}	4	36
Othello	10^{28}	10^{58}	10	58
Checkers	10^{21}	10^{31}	-	-
Chess	10^{46}	10^{123}	35	80
Shogi	10^{71}	10^{226}	92	115
Go	10^{72}	10^{360}	250	150

tree, specifying Min and Max's playing turn, and gives scores to the leaf nodes based on the evaluation of some heuristic function. It then backtracks Max's winning path consisting of a series of moves.[7]

The tree structure is an abstract mathematical concept for representing objects and their relationships. The tree consisting of nodes and branches resembles a natural tree turned upside down (Figure 5.1). Two nodes in the tree have no more than one path between them. The starting node is called the root, and the end nodes are called leaves. A node is called a parent node if it has children nodes connected to it. Multiple nodes proceeding from the same parent are called siblings. The arc or edge between two nodes is called a branch.

The Min-Max algorithm with its variants was a classic game-playing algorithm used in the early days of AI before the advent of reinforcement learning. Generating the entire game tree for any of the board games shown in Table 5.3 is practically impossible, given the size of the datasets. To overcome this problem, Min-Max adopts various techniques to prune the branches of the tree.

The method at best is heuristic and not exact. The biggest disadvantage of this game-playing algorithm is that there is no learning. Therefore, the agent has no game-playing policy to make the winning moves. Latest research has repeatedly demonstrated that the best way of circumventing the problem of generating unrealistic massive datasets is to embed some kind of reinforcement learning in the system.

DATA AUGMENTATION

Deep learning models consume tons of data, just not in quantity but also in diversity.

However, it is very difficult to increase the instances in datasets for learning. Data augmentation refers to the process of making new data instances by making minor modifications to the available data instances without introducing new and fresh instances from external sources. Most of the deep learning frameworks contain built-in data augmentation utilities.

By improving generalization, data augmentation leads to an overall prediction accuracy. It also helps overcome overfitting, a phenomenon in which the network learns a function that perfectly fits the training data.

Image data augmentation

Consumer goods recommendation systems use a lot of pictures and images. In addition, enterprises dealing with fashion, travel, apparel, real estate, photography, and so on cannot do their business without handling pictures. Knowledge of object recognition and image processing algorithms along with the standard image datasets is a must for such enterprises.

Convolutional neural networks (CNNs) are state-of-the-art deep learning networks for computer vision tasks like object recognition, detection, semantic segmentation, and so on. Some of the well-known publicly available datasets to test the cutting-edge CNN algorithms are CIFAR-10, CIFAR-100, ImageNet, MNIST (hand-written digit recognition), SVHN (street view house numbers), PASCAL VOC12, COCO, and so on (refer to Appendix A). Massive as they are, researchers claim that these datasets still do not contain sufficient instances to raise the prediction accuracy of their algorithms. Numerous data augmentation techniques are tried and tested on the above datasets. Some of the traditional data augmentation techniques involve horizontal/vertical flipping, rotation, cropping, saturating, blurring, adding Gaussian noise, and so on (Figure 5.2).

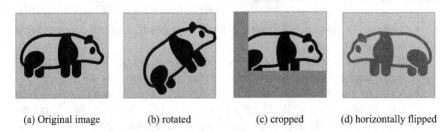

(a) Original image   (b) rotated   (c) cropped   (d) horizontally flipped

Figure 5.2 Image data augmentation.

The dataset extended by augmentation increases by a factor equal to the number of transformations applied. Since the data augmentation is a random process, there is no guarantee that the ML algorithm will produce a proportionate increase in the prediction accuracy. Besides, care should be taken when applying the transforms to a dataset. For example, a rotation or a flip of an image in the ImageNet dataset will result in a new data instance, but it cannot be done for digit images like 6 or 9 in the MNIST hand-written digit dataset.

Mixing images together by averaging their pixel values, although quite counterintuitive, is also considered to be an effective augmentation strategy.[8] AutoAugment is an algorithm proposed by Google to learn the best augmentation policies for a given dataset with the help of reinforcement learning. A policy comprises five subpolicies, and each subpolicy comprises two image operations applied in sequence. Each image operation in turn considers two parameters: the probability of calling the operation, and the magnitude of the operation. If the randomly chosen probability of calling an operation turns out to be null, then the operation may not be applied in that mini-batch. However, if applied, it is applied with a fixed magnitude.[9]

In practice, AutoAugment algorithm is computationally expensive. Population-based augmentation (PBA) is an alternative data augmentation technique with less compute. It generates nonstationary augmentation policy schedules instead of a fixed augmentation policy.[10]

Numerical data augmentation

The above section described some of the statistical techniques for image data augmentation. The techniques are fairly straightforward for image data. Individual images are randomly cropped, rotated, sprinkled with random noise, and color contrasted to increase the number of available samples. The augmentation techniques work well in image datasets because minor perturbations at the pixel level do change the image class. Unfortunately, such a method does not work for numerical datasets. A tiny perturbation in the numerical data can shift the perturbed sample into an entirely different class. The following simple algorithm[11] is effective in producing data augmentation in numerical data:

- Copy each class in a subset a couple of times to achieve a reasonable representation.
- Randomly perturb features in each subset using the distribution's mean and standard deviation as perturbation bounds.
- Validate that each new sample belongs to a proper class or reassign to a new class using Mahalanobis Distance (MD) and Gaussian Mixture Models. MD is a multidimensional generalization of how many standard deviations away a sample is from the mean of a distribution.

Text data augmentation

The application of ML to Natural Language Processing (NLP) tasks, too, requires a very large collection of text data. Although text data is abundant and freely available in the form of web pages and other digital repositories, preparing the data with annotations and in formats amenable for ML is an extremely tedious and expensive task. Data augmentation is not very common in NLP tasks, but some significant amount of research is already in progress. Word-level text data augmentation and sentence-level text data augmentation described below are the two major approaches proposed.

Word-level text data augmentation

This method, also known as lexical substitution, tries to substitute a randomly chosen word in a sentence either with a synonym or another word not too far in meaning. Synonyms are readily available in WorNet,[12] which is a standard digital thesaurus used in NLP tasks. Typical similarity metrics used are k-nearest neighbors and cosine similarity. Yet another method is to substitute words from the pretrained word embeddings such as Word2Vec[13] of FastText.[14] By far, the most successful method is the use of transformer models developed by Google such as bidirectional encoder representations from transformers (BERT).[15] BERT is pretrained on a large body of text using masked language modeling in which the model has learned to predict masked words based on the context. Hence, in text data augmentation, BERT can replace a masked word which does not alter the meaning of the original sentence. A further improvement on word embedding data augmentation is the contextual augmentation method which offers a wider range of substitute words predicted by a bidirectional language model matching the context[16]

Sentence-level text data augmentation

Sentence shuffling in short paragraphs of text is also known to add useful data instances to a training body of text. Python NLTK library offers sentence tokenization and shuffling. This is useful in sentiment analysis studies, because with sentence tokenization, the sentiment conveyed in the original text is still maintained. Paraphrasing is another useful method. BERT is quite powerful in paraphrasing a phrase or a sentence owing to the masking method. Neural machine translation is an upcoming technology in the field of translation. It needs a tremendous amount of data in the form of parallel corpora of the source and the target languages. There is not always a 1-1 correspondence between the words in the two languages. Words in the target language which are not present or omitted in the source language are introduced in the source language by means of back-translation. Effectively, back-translation serves as an instrument to augment data in a target translation language, which may not have sufficient data for training a machine translation model.[17,18]

Synthetic dataset

Even the exquisite data augmentation techniques described above do not suffice to produce the high quality and the right amount of data necessary for accurate machine learning. Another well-known method is creating a repository of synthetic data algorithmically. The main purpose of this technique is to generate flexible and rich data to experiment with various regression, classification, clustering, and deep learning algorithms.

Privacy concern of customers is another notable reason for generating synthetic data. For example, predicting customer behavior or preferences would need large amounts of data to build the learning models. But because of privacy, accessing real customer data is slow and does not provide good enough data on account of extensive masking and reaction of information.[19] In such cases, generating synthetic data from a model fitted with real data is the need of the hour. If the model is a good representation of the real data, then the synthetic data will also pick up the statistical properties of the data, which are indispensable for learning.

Table 5.4 shows some of the relevant properties a synthetic dataset should possess.

Examples of synthetic datasets include SURREAL[20] for estimating human pose, shape, and motion from images and videos; SYNTHIA[21] for semantic segmentation of urban scenes, and those for text detection in natural images.[22]

DYNAMIC LEARNING FRAMEWORK

This section describes a novel dynamic learning framework that partially automates the ML pipeline. This unit of the framework contains the usual four categories of ML: supervised, unsupervised, semi-supervised, and reinforcement learning. Each ML category contains more algorithms than those described in Chapter 4. The framework consists of several computational units that work together seamlessly. Eventually, dynamic learning can be implemented by DL bots that autonomously decide on the type of

Table 5.4 Dataset format

Data attributes	Desired properties
Type	It can be numerical, binary, or categorical.
Features	The number of features of the dataset should be flexible.
Size	The size of the dataset (number of records) should be flexible.
Distribution	The dataset should be based on a statistical distribution, which can be varied.
Noise	A controllable amount of random noise should be injected in the dataset.

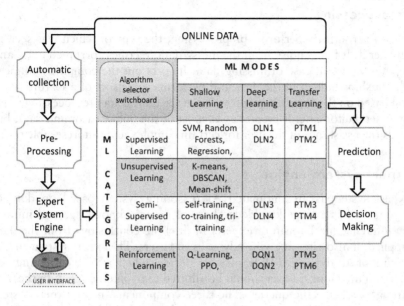

Figure 5.3 Online dynamic learning framework.

ML models to be used. Bots can learn dynamically from previous decisions and can also self-heal. DL systems can anticipate customer needs, market changes, and even cybersecurity threats. Contextual, personalized, and secure services can be provided by dynamic learning framework. Figure 5.3 outlines this dynamic learning framework.

The frontend of the ML framework contains the following subunits:

Online data repository

The customer data or any other data that is of interest to an enterprise is stored in an online repository or cloud. This data, being dynamic in nature, is constantly changing. The concerned enterprise may or may not be directly in control of the data sources responsible for stacking and managing the data in the online repository.

Automatic collection

The automatic data collection unit constantly monitors the ever-changing data in the online repository. Whenever new chunks of data become available, it automatically downloads it and pushes it to the preprocessing module. The automatic collection module is responsible for ascertaining that there is no data redundancy.

Preprocessing

This unit routinely performs preprocessing, the type of which was seen in Chapter 4. It filters noise from the messy data, rectifies corrupted data, and imputes values in case of missing data. It sorts out the complex nonlinear relationships hidden in the data and performs some sort of a feature selection based on the expert rules provided in the unit. Data preprocessing module is semiautomatic. It preprocesses most of the online incoming data, but needs the assistance of the user for certain complex and unforeseeable tasks.

Expert system engine

Expert system is an automated AI reasoning system based on domain experts. It has a large knowledge base of domain knowledge and mimics the inference mechanism of experts in drawing conclusions from the data supplied. It also aims at giving advice to end users. The expert system engine subunit of the dynamic ML framework decides which machine learning category an incoming dataset should be allotted to by interacting with the user through data-specific queries. The three components in the expert system engine are knowledge acquisition, knowledge representation, and inference (Figure 5.4). These are described in the following subsections.

Knowledge acquisition

This is the very first step in building an expert system. Knowledge predominantly resides with the domain experts. In the ML context, data scientists are

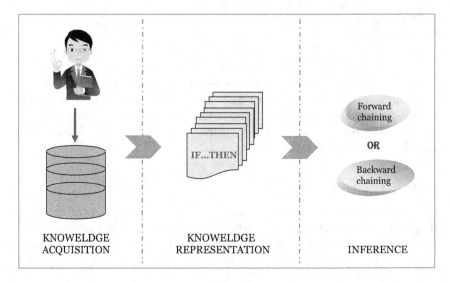

Figure 5.4 Constituent elements of expert system engine.

the domain experts who have detailed knowledge of data and the associated ML methods and algorithms. In particular, they know which ML category or algorithm a particular dataset should be directed to. The body of knowledge is systematically collected from data scientists and stored in a convenient format in a large knowledge repository called the knowledge base.

Knowledge representation

The knowledge acquired from the expert may be in the form of unstructured text, tables, diagrams, and annotations. This knowledge is then represented in a well-defined format to remove ambiguities and make the coding work easier. Semantic networks, frames, and ontologies are some of the well-known knowledge representation formats in the knowledge engineering domain. The IF-THEN rules are then framed from the knowledge formats. For instance, there will be rules like IF (data contains input–output pairs) THEN (supervised learning), IF (data does not contain input–output pairs) THEN (unsupervised learning), etc. The framework contains a user interface at the front end. This is for the expert system engine to query with the user about the characteristics and type of online incoming data.

Inference

Inference is the process of drawing conclusions by linking data precepts. The main inference mechanisms found in modern expert systems are forward-chaining, backward-chaining, and hybrid. In the forward-chaining or data-driven inference mechanism, if a piece of data matches the "IF" part of any "IF-THEN" production rule, the rule fires and produces another relevant piece of data. The newly produced intermediate conclusions of all the rules that have fired are kept in the working memory of the expert system. The inference engine then searches for rules in the ruleset that match the new contents available in the working memory. These rules are then fired, generating more intermediate conclusions which are used to invoke applicable rules in the next round. The fetch-match-fire cycle continues till the inference engine exhausts all the fire-able rules that exist on the agenda. In the backward chaining or goal-driven inference mechanism, the "THEN" part of the "IF-THEN" rule is provisionally assumed to be true, and the inference engine looks for data that will satisfy this subgoal. It then seeks to link all the subgoals to yield the final goal. The hybrid inference is a combination of the forward and backward chaining. Given a dataset and a ruleset, the expert system reaches the same conclusion irrespective of the inference chaining. The only difference is the speed of convergence, which is problem dependent.

In a medical diagnosis session, for example, the expert system initiates the session by asking the patient a couple of questions related to the patient's symptoms through the user interface. For instance, a backward chaining inference expert system, which is quicker than the forward chaining inference system, will hypothesize the patient has a particular sickness based on the initial list of symptoms the patient has supplied and then proceed to get further information (data) from the patient to prove the hypothesis. A similar strategy is employed in the inference mechanism of the dynamic ML expert system engine. It will first query with the user about the specifications of the data as shown in Table 5.2 and accordingly decide on the *category* and *mode* of learning.

ML MODES IN DYNAMIC LEARNING

Shallow learning

The term "shallow learning" is a misnomer because the performance of the ML algorithms pertaining to this class is far from being shallow. "Shallowness" is not a salient feature of the "shallow learning" ML algorithms. The term is collectively and rather loosely used to refer to all forms of ML before the advent of "deep learning" (Figure 5.5).

The common feature of the algorithms which are included in "shallow learning" is that they rely on handcrafted features based upon heuristics of the target problem. The domain experts and data scientists define the features of the problem space, and subsequently, data is collected in a format dictated by the features. Data is preprocessed and cleaned, and then the features are further refined. Finally, the cleaned data is fed into algorithms like linear regression, SVM, random forests, and neural networks (NNs) to arrive at prediction and decision-making thereof. Viewed this way, it is closer to truth to call shallow learning as feature engineering.

NN is one of the most common ML techniques found in the AI research community. Historically, neural networkswere proposed in the 1940s as

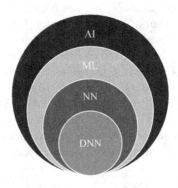

Figure 5.5 AI, ML, NN, and DNN.

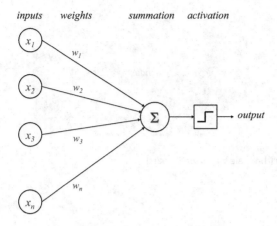

Figure 5.6 Rosenblat's perceptron.

perceptrons (Figure 5.6).[23] They were found to be capable of solving several problems of interest at the time. However, research in the potential applications soon began to decline when some researchers criticized NN as not being able to perform the XOR operation. The problem was solved years later when a middle layer was introduced in the NN. It is precisely the number of layers in a NN that distinguish deep learning from shallow learning.

Corresponding to each category and mode of learning, there are specific algorithms that operate on the incoming data. The shallow learning mode, for example, contains all the algorithms described in Chapter 4 and a few more relevant ones that are not covered in this book.

The algorithm selector switchboard (Figure 5.7) in the framework helps the user to select an appropriate ML algorithm for an ML training session. The system will use an algorithm indicated by the user. In the ensemble mode, it will use an ensemble of algorithms and report the best performance. When operating in the default setting, the system will choose an algorithm at random from multiple algorithms to perform learning.

Deep learning

Deep learning methods have gained popularity because they often outperform conventional (i.e., shallow) ML methods and can extract features automatically from raw data with little or no preprocessing.[24,25,26] The structure of a deep neural network is very complex, containing layers in the order of tens or hundreds. The VGGNet,[27] which is often used as a pretrained artifact for image recognition, consists of 16 convolutional layers. As opposed to the shallow neural networks that do a mere weighted summation of the inputs at each of the neurons, deep CNNs perform complex operations like weight sharing, convolution, and pooling during each forward pass.

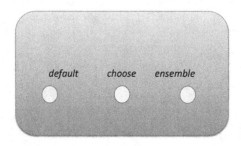

Figure 5.7 Algorithm selector switchboard.

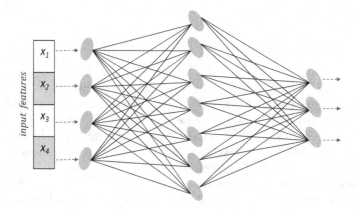

Figure 5.8 Shallow neural network.

The essential difference between shallow learning (feature engineering) and deep learning (feature learning) is illustrated in Figures 5.8 and 5.9. In shallow learning, the features are handcrafted and fed into NN, which has only one hidden layer (and, therefore, "shallow"). In contrast, the deep network has a large number of hidden layers, making the structure "deep." Handcrafted features are not fed into the Deep Net. Instead, these hidden layers successively learn the features (Figure 5.9). Deep learning, therefore, is also referred to as feature learning.

Classification problems involving text and images cannot be handled by shallow learning. Deep convolutional neural networks and deep recurrent neural networks are best suited for these learning tasks. In the deep learning mode of our framework (Figure 5.3), there are a set of deep learning models DLM1, DLM2, ...DLMn, from which the most suitable one is selected by the framework to engage in the learning task.

Machine learning, in general, has made remarkable progress in recent years. But the outstanding success is achieved in the area of DL. Deep

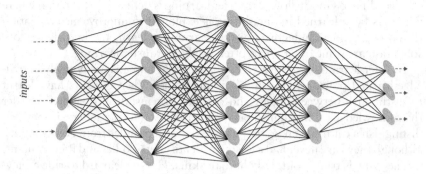

Figure 5.9 Deep neural network.

learning systems now enable previously impossible smart applications, rev-
olutionizing image recognition and natural language processing, and identifying complex patterns in data.[28] Deep learning has shown great promise
for tackling many tasks such as image processing,[29] natural language processing,[30] speech recognition,[31] superhuman game playing,[32] and autonomous driving.[33,34]

Transfer learning

In this section, there are several pretrained models, each with a specific kind
of input–output data. These are chosen by the system depending on the new
incoming data.

ML comprises two distinct phases: training phase and testing phase. One
of the fundamental assumptions of ML is that the training and the testing
datasets come from identical statistical distributions. To a large extent this
is justified because in most of the ML cases, only one original dataset is
dealt with. This original dataset is randomly divided into training dataset
and testing dataset. The ML experiment is repeated to yield a statistically
satisfiable k-fold cross-validation, which on average leads to a minimization
of the loss function in training and minimization of the prediction error in
testing. However, the power of ML is demonstrated in the real world when it
is trained on a *known* (historical) dataset and then used to make predictions
on *unknown* datasets – future datasets that were not available at the time of
training. This is where most of the ML models get into trouble. The future
test datasets may not share the same statistical distribution as the parent
training dataset. The performance will degrade. The most natural thing to
do when faced with such a situation is to rebuild another ML model from
scratch to fit the new data and start the training-validation-testing phases all
over again. This is where transfer learning comes to help.

According to Goodfellow,[35] transfer learning is defined as "Situation where what has been learned in one setting is exploited to improve generalization in another setting." Inductive transfer, learning to learn, and knowledge consolidation are some other terms often used for transfer learning.

Transfer learning is motivated by the fact that humans easily apply knowledge learned previously to solve new problems. When toddlers have learnt to distinguish between apples and oranges, for instance, they can transfer their knowledge to distinguish between bananas and cucumbers; if they can distinguish a cat from a dog, they can distinguish poodles from Pomeranians, although they have never had a lesson in identifying poodles and Pomeranians.

The same is true about daily human skills. Having learned to ride a bicycle, intuitively basic bicycle-riding skills like balance and maneuvring are transferred to ride a motorbike. Driving on the left side of the road with right-side steering is easy to adapt to driving on the right side of the road with left-handed steering. Similarly, for an agent trained to drive a vehicle on a 2D surface, transfer learning extends the agent's capabilities to drive in 3D space. Transfer learning is used to further train computer programs that have learnt autonomous driving of land vehicles to become drone drivers. The programs will tacitly use the knowledge they have picked up in dodging obstacles while driving on the land to avoid obstacles in mid-air. Thus, transfer learning is the ability to transfer knowledge across tasks. The more related the tasks, the easier it is to transfer knowledge across the tasks.

ML AUTOMATION AND OPTIMIZATION

The dynamic ML framework proposed in this chapter has several benefits: It is simpler to use for nonexperts, more reliable, and faster to deploy and yields better performance than hand-designed models. There are three major issues concerning the functionality of the framework. (a) How will the system select the *ML category*, the row in the framework? (b) How will the system select the *ML mode*, the column in the framework? (c) How will the system deliver an efficient performance?

The answer to the first two questions is *automation*. The answer to the third question is *optimization*.

In selecting the ML category (rows in the framework), the system is automated to switch among the following three strategies:

1. User selection using the switching board (Figure 5.7)
   A user experienced in data science or machine learning can select any row or column, as well as any ML category or mode of learning provided in the framework to perform the learning task. If the user's choice is appropriate to the learning task with the data provided at the input, the system will proceed smoothly. However, it will stop with an

error code if the selected category or mode of learning does not match the input data.

2. Data-driven strategy

The system will gauge the scale of the data and select an appropriate algorithm in the given category. For simple and small datasets, it will select simpler algorithms; for larger and more complex datasets, it will select more complex algorithms from the suite provided in the framework.

3. Ensemble Learning strategy

For a given learning problem, the framework will try multiple algorithms and select the one with the best performance. The disadvantage is computational time. However, it is ameliorated since the dynamic ML framework is operating online with continuous self-monitoring. Besides, modern enterprises are equipped with massive and rapid computing resources spread across the cloud. Graphic processing unit (GPU) computing which offers orders-of-magnitude performance increase over the conventional CPU computing is an additional blessing.

ML is an optimization problem. Most ML algorithms salvage through heaps of data to learn a mapping function from the inputs to the output or learn to cluster the data based on some implicit function optimization. NN has become the de facto standard of ML. As a matter of fact, the concept of deep learning has emerged directly from the application of NN for ML tasks. The NN used for shallow learning has a single hidden layer sandwiched between the input and output layers. As more and more hidden layers are stacked to handle complex problems and improve the fine granularity of learning, the NN structure becomes deeper and deeper culminating in the current paradigm of deep learning.

NN has three classes of parameters: the network topology (number of layers and number of neurons in each layer), the connection weights, and hyperparameters. The NN connection weights are trained, validated, and tested by using the training, validating, and testing data subsets, respectively. The randomly initialized connection weights are gradually optimized by the gradient-based backpropagation algorithm through the epochs of training. It is evident that in the current ML procedures, only the connection weights of the NN model are optimized. NN topology and hyperparameters are not systematically optimized. Although there is a popular dropout technique to alter the topology of the network while training, data scientists manually adjust the number of layers along with the number of neurons in each layer and tweak the values of the hyperparameters with the hope of improving the performance of the ML algorithm. The trial-and-error method is far from being efficient.

A far more efficient way that combines ML with evolutionary algorithms is neuro-evolution, described in the following subsection. This strategy simultaneously optimizes the topology of the network and the hyperparameters while optimizing the connection weights.

Neuro-evolution

Neuro-evolution is a relatively new paradigm that combines optimization and evolution in the context of the NN training. The optimization problem is framed in a rigid mathematical set of equations, but the process of optimization follows a flexible nature or bio-inspired computational metaphor, which is often referred to as soft computing.[36] Neuro-evolution,[37] in particular, adapts an evolutionary algorithm approach.

Evolutionary algorithms are a successful area of AI. These algorithms are intuitive and easy to program. Their great advantages are that they are robust and domain-independent. This means they depend neither on the problem size nor on the application domain. With minor modifications, they can be used as black boxes to optimize any objective function in any domain, provided the problem can be framed as an optimization problem illustrated in the next subsection.

Neuro-evolution integrates the well-known genetic algorithm (GA), which is a computational metaphor of the Darwinian paradigm of natural selection. The optimization problem formulation and the GA[38] are dealt with in the following subsections:

Optimization problem formulation

An optimization problem is defined as:

Find X such that $f(X)$ is minimum/maximum $\qquad\qquad$ (5.1)

subject to the constraints:

$$g(X) = 0 \qquad\qquad (5.2)$$
$$h(X) < 0 \qquad\qquad (5.3)$$

where X is a vector of decision variables, usually bounded:

$$X_{min} \leq X \leq X_{max} \qquad\qquad (5.4)$$

$f(X)$ is the objective function, $g(X)$ and $h(X)$ are the constraints, and X_{min} and X_{max} are the bounds on the decision variables in vector X.

Genetic algorithm

Figure 5.10 shows the flowchart of the GA, which is inspired by the natural selection process. It randomly generates feasible solutions called parents to an optimization problem and computes their fitness based on the value of

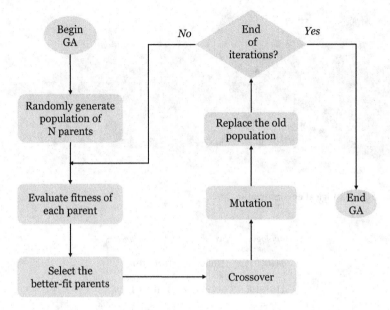

Figure 5.10 Genetic algorithm flowchart.

the objective function. These parents are then allowed to evolve through a series of evolutionary cycles consisting of various genetic operators. Over the course of evolution, the selection of fitter parents in the population and the eventual crossover and mutations in the successive generation of off-spring gradually result in better fit or optimal individuals.

The GA iterates through the following cycles:

Step 1 – Random generation of population

A population of N individual solutions is generated randomly. Care must be taken to ensure that the individuals do not violate the constraints imposed on the input variables. The individual solutions that make up the population are called parents or chromosomes. The bit representation of these solutions resembles a biological chromosome composed of a long strand of bases as shown in Figure 5.11.

Each individual solution is a vector X containing the values of the variables. The neuro-evolution problem deals with the simultaneous evolution of the neural network topology, the connection weights, and the hyperparameters. In the binary bit string representation, each parent or chromosome (solution) is a band containing three 8-bit units for topology, weights, and hyperparameters (Figure 5.12).

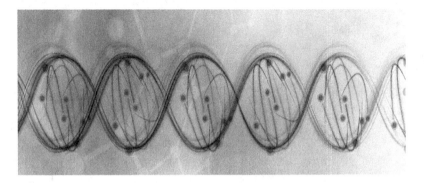

Figure 5.11 Biological chromosome.

topology	weights	hyperparameters
11100011	11001110	10100101
11010100	10101010	10101101

Figure 5.12 Pair of GA chromosomes encoded as bit strings.

Step 2 – Fitness evaluation

The fitness function $f(X)$ is evaluated. In most cases, the evaluation of $f(X)$ is a direct computation. However, in practical application areas, the evaluation of $f(X)$ may involve a time-consuming elaborate simulation. The fitness function for maximization problems is directly proportional to the value of the objective function. For minimization problems, it is customary to consider the reciprocal of the objective function as the fitness. In ML, some derivative of the loss function or the reward function may be used as the GA fitness.

Step 3 – Selection

The greater the value of the fitness function, the fitter the individual. The better-fit parents in the population are selected for crossover (also called recombination). There are two main types of selection schemes: tournament and roulette wheel. In the tournament selection, a pair of parents is selected at random from the population. Their fitness is compared and the fitter of the two is selected for "reproduction." In case of a tie in fitness values, the selection is performed randomly. The selection procedure is repeated till the number of the selected parents equals the population size. In the roulette wheel selection scheme, the chromosomes are treated as if they are placed on a roulette wheel according to their fitness. The chromosomes that occupy the greater area on the wheel have a greater chance of being selected. For each selection, the roulette wheel is rotated as in a casino, and the chromosome to which the pointer points to when the roulette wheel stops is selected.

Step 4 – Crossover

In the natural world, crossover mixes the genetic material in the offspring of the species and increases its chances of survival. The following three kinds of crossover operators are common in GA.

1. One-point crossover: A single crossover point on both the parents' strings is randomly selected. The part of the chromosome after the crossover point is swapped between the two parent organisms (Figure 5.13a).
2. Two-point crossover: Two distinct points are selected on the parent chromosome strings. The part of the chromosome between the two crossover points is swapped between the parent organisms (Figure 5.13b).
3. Uniform crossover: Each corresponding bit between the parent chromosomes is swapped with a small probability p (Figure 5.13c).

Step 5 – Mutation

Every bit in every individual is flipped (0/1) with a very small mutation probability. Mutation makes the search wider and aids premature convergence of the population.

Step 6 – Inserting the new offspring in the older generation

Since the initial solutions are randomly generated and evolved through the GA operators, there is no guarantee that the final solutions will always remain within the bounds imposed by the problem constraints. Any infeasible solutions are "repaired" and then inserted back to the old population, and the above steps of the GA cycle are iterated.

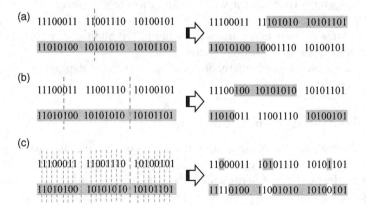

Figure 5.13 (a) One-point crossover (b) Two-point crossover (c) Uniform crossover.

The dynamic ML framework handles all the above steps automatically. Once the learning category and mode are selected, the system automatically rolls the GA in motion to optimize the learning model performance.

Recommendation systems

Deep learning techniques may be used to develop a realistic recommendation system for online shopping customers.

Almost all businesses have gone online and keep a record of the products purchased by customers along with the customer preferences and their personal data. Further, to enhance their sales, online businesses are routinely engaged in maintaining recommendation systems. These systems, which monitor customer preferences along with their data, quickly run an ML algorithm over the collected data and come with a recommendation of new products which may interest the customers. The recommendations are not totally unrelated random suggestions, but are based on the customer purchase history and their preferences. In a nutshell, recommendation systems are ML systems that learn to predict the rating or preference an online customer would give to an item or product in the market.

When a customer gets in a physical store (as opposed to an online store), she can see all the items displayed in the store. Since the collection of items is not too large, she can easily select a handful of items to focus on at a time, and then make her purchase choice. However, this kind of a stress-free traditional buying suddenly transforms into a confusion and stressful situation when the customer takes her shopping spree to online stores. First of all, she has to search for relevant items using a web-based search engine; she is immediately bombarded with a vast collection of items, and it becomes extremely difficult to make the buying choice. Recommendation systems come to her rescue by suggesting a relevant list of items, not too large, so that she can concentrate to make her selection. Recommender systems are information filtering systems that offer a win-win situation to the buyer and the seller. It helps the buyer to select a handful of items at a time to make her buying choice and, simultaneously, helps the seller to sell additional products.

Types of recommendation systems

Below, different kinds of recommendation systems deployed in practice are explained.

Popularity-based method

This is the simplest recommendation system conceivable. It is based on the popularity of the item being sold. The popularity of an item is simply the count of the number of items sold so far. Items in the online store are given a popularity index directly based on the number of pieces of that item or product being sold, and then ranked according to this popularity index.

The most popular items are placed on top of the list recommended to the user. The major drawback of the popularity-based method is that it does not offer any *personalization* to the customer.

Collaborative filtering

It has become a common practice for online marketing enterprises to get the customers to rate the products in a casual and noninvasive way when they are busy inspecting the items on display in the online store before making a purchase. The ratings dataset is converted into a matrix in which customers occupy the rows and items occupy the columns. The cell corresponding to a customer and an item is the rating given by the customer to that particular item, on a scale of 1~5 in this case (Figure 5.14). The interaction matrices are usually very large, although sparse.

Collaborative methods work by employing appropriate machine learning algorithms that try to learn a function that predicts the preference of items to each user. This is based on a similarity measure that computes the similarity of users with respect to the items.

The *k*-nearest algorithm often used in recommendation systems goes through the following steps:

- Compute the similarity of users using cosine or Pearson similarity measure
- Determine *k* users close in similarity to the user *u* in question
- Recommend items preferred by k users, but not yet purchased by user *u*.

	item1	item2	item3	item4	item5
user1	1	4		3	1
user2	2	5			2
user3	1	4	4	1	
user4	2		3	2	2
user5		5	2		1

Figure 5.14 Users-items interaction matrix.

Deep learning for recommendation systems

Recently, the freely available Web video and music channel YouTube has become immensely popular among its users. The Web app displays video clips ready for playing on demand. The keyword search also is appealing and effective. One of the most outstanding features of YouTube is the recommendation mechanism that comes hand in hand with the services. For example, when searching for a piece of classical music composed by Vivaldi, it will instantly come up with a list of Vivaldi's violin sonatas; and depending on what the user chooses to play, the list of recommendation will modify itself in real time to provide the user with the best possible choice. It will also instantly recommend related Baroque composers.

These systems deal with rapidly changing large sources of data. Millions of video and music clips are uploaded on the internet on a daily basis. Conventional ML system based on shallow learning cannot cope up with the scale and complexity of such recommendation systems. Deep learning is the best-fit candidate.[39] (Figure 5.15)

Figure 5.15 shows the general architecture. The central unit of the system contains two prominent deep neural networks (DNN). The first DNN is fed with the online datasets and learns to perform collaborative filtering based on features such as user's age, gender, history of search and watch, related genre, and so on. From the purchasing history of similar customers, it learns to make predictions (filtering) about the interests of the customer in question. The outcome of the collaborative filtering DNN is a filtered subset of media (music or videos). The filtered subset of media is then fed to the second DNN along with the video features (genre, stars, age, bag of words from lyrics, etc.). The second DNN fine-tunes the selection according to users' taste and produces a still smaller subset of the most relevant media to the user. These are ranked and then displayed in the rank order.

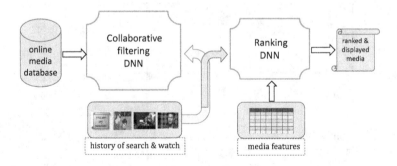

Figure 5.15 Deep learning recommendation system (adapted from Covington et al.)

Data for fuelling recommendation systems

Large-scale enterprises do not have any dearth of data which they readily employ to drive several ML systems. Medium-sized enterprises also have sufficient data which can run the online ML engines. However, small-size enterprises and start-ups often struggle to reach the data critical mass that will make their ML algorithms effective. A prominent pain point here is how to quickly amass data that will drive the product recommendation systems that those firms have freshly deployed. A viable alternative is to experiment with datasets freely available in the public domain such as Kaggle.[40]

CONSOLIDATION WORKSHOP

1. What are the benefits of static learning? What are its limitations?
2. Describe the characteristics of dynamic learning?
3. What are some of the most important data characteristics?
4. What is the importance of static data?
5. Describe the differences between preprocessed and distributed data.
6. What are the challenges of continuously changing data? How is it monitored?
7. What are the characteristics of shallow learning?
8. What is the impact of deep learning?
9. What are the challenges of transfer learning? Give some concrete examples where one can apply transfer learning.
10. How is shallow learning used in data analytics?
11. How is deep learning used in data analytics?
12. What is neuro-evolution? How can it be applied to ML for BO?
13. How are recommendation systems implemented in businesses?
14. What is the role of the discount factor in deep Q learning?
15. What is a cold start in recommendation systems? How is it resolved?

NOTES

1. Pykes, Kurtis, Getting a data science job is harder than ever: How to use the difficulties of landing a gig to your advantage. *Medium*, 2020, September 29, https://towardsdatascience.com/getting-a-data-science-job-is-harder-than-ever-fb796aae1922.
2. Davenport, Thomas H. and Patil, D.J., Data scientist: The sexiest job of the 21st century. *Harvard Business Review*, 2012, October. Issue https://hbr.org/2012/10/data-scientist-the-sexiest-job-of-the-21st-century.
3. https://developers.google.com/machine-learning/crash-course/static-vs-dynamic-training/video-lecture.

4. Cortez, P., Cerdeira, A., Almeida, F., Matos, T. and Reis, J., Modeling wine preferences by data mining from physicochemical properties. *Decision Support Systems*, Elsevier, 47(4): 547–553, 2009.
5. https://archive.ics.uci.edu/ml/datasets/wine+quality.
6. Kumar, Aviral, Fu, Justin, Tucker, George, and Levine, Sergey, Stabilizing off-policy Q-learning via bootstrapping error reduction. *Advances in Neural Information Processing Systems*, 32, NeurIPS 2019.
7. Millington, Ian, *AI for Games*, CRC Press New York, NY; 3rd ed. 2019, March 26.
8. Inoue, Hiroshi, Data augmentation by pairing samples for images classification. *ArXiv e-prints*, New York, NY, 2018.
9. Cubuk, E. D., Zoph, B., Mané, D., Vasudevan, V., and Le, Q. V., AutoAugment: Learning augmentation strategies from data. *2019 IEEE/CVF Conference on Computer Vision and Pattern Recognition (CVPR)*, Long Beach, CA, pp. 113–123, 2019. Doi: 10.1109/CVPR.2019.00020.
10. Ho, D., Liang, E., Chen, X., Stoica, I., and Abbeel, P., Population based augmentation: Efficient learning of augmentation policy schedules. In *International Conference on Machine Learning*, pp. 2731–2741. PMLR, 2019, May.
11. https://towardsdatascience.com/augmenting-categorical-datasets-with-synthetic-data-for-machine-learning-a25095d6d7c8.
12. Fellbaum, C., WordNet. In Roberto Poli, Michael Healy, & Achilles Kameas (eds.), *Theory and Applications of Ontology: Computer Applications*, pp. 231–243. Springer, Dordrecht, 2010.
13. Goldberg, Y., and Levy, O., word2vec Explained: Deriving Mikolov et al.'s negative-sampling word-embedding method. *arXiv preprint arXiv:1402.3722*, New York, NY, 2014.
14. Joulin, A., Grave, E., Bojanowski, P., Douze, M., Jégou, H., and Mikolov, T., Fasttext. zip: Compressing text classification models. *arXiv preprint arXiv:1612.03651*, 2016.
15. Devlin, J., Chang, M. W., Lee, K., and Toutanova, K., Bert: Pre-training of deep bidirectional transformers for language understanding. *arXiv preprint arXiv:1810.04805*, 2018.
16. Kobayashi, S., Contextual augmentation: Data augmentation by words with paradigmatic relations. In *Proceedings of the 2018 Conference of the North American Chapter of the Association for Computational Linguistics: Human Language Technologies*, Volume 2 (Short Papers), pp. 452–457, 2018, June.
17. Sugiyama, Amane, and Yoshinaga, Naoki, Data augmentation using back-translation for context-aware neural machine translation. In *Proceedings of the Fourth Workshop on Discourse in Machine Translation* (DiscoMT 2019), Hong Kong, China, pp. 35–44, 2019, November 3.
18. Sennrich, Rico, Haddow, Barry, and Birch, Alexandra, Neural machine translation of rare words with subword units. In *Proceedings of the 54th Annual Meeting of the Association for Computational Linguistics (ACL)*, Berlin, DE, pp. 1715–1725, 2016.
19. El Emam, Khaled, Mosquera, Lucy, and Hoptroff, Richard, *Practical Synthetic Data Generation: Balancing Privacy and the Broad Availability of Data*, O'Reilly Media Sebastopol, CA; 1st ed., 2020, May 19.

20. Varol, G., Romero, J., Martin, X., Mahmood, N., Black, M. J., Laptev, I., and Schmid, C., Learning from synthetic humans. In *Proceedings of the IEEE Conference on Computer Vision and Pattern Recognition*, Honolulu, HI, pp. 109–117, 2017.

21. Ros, G., Sellart, L., Materzynska, J., Vazquez, D., and Lopez, A. M., The synthia dataset: A large collection of synthetic images for semantic segmentation of urban scenes. In *Proceedings of the IEEE Conference on Computer Vision and Pattern Recognition*, Las Vegas, NV, 3234–3243, 2016.

22. Gupta, Ankush, Vedaldi, Andrea, and Zisserman, Andrew, Synthetic data for text localisation in natural images. In *Proceedings of the IEEE Conference on Computer Vision and Pattern Recognition (CVPR)*, Las Vegas, NV, pp. 2315–2324, 2016.

23. Rosenblatt, Frank, *The Perceptron—A Perceiving and Recognizing Automaton*. Report 85-460-1. Cornell Aeronautical Laboratory, Cornell, NY, 1957.

24. Liang, Hong, Sun, Xiao, Yunlei, Sun, and Gao, Yuan, Text feature extraction based on deep learning: A review. EURASIP *Journal on Wireless Communications and Networking*, 2017, Doi: 10.1186/s13638-017-0993-1.

25. LeCun, Y., Bengio, Y., and Hinton, G., Deep learning. *Nature*, 521: 436–444, 2015, Doi: 10.1038/nature14539.

26. Krizhevsky, A., Sutskever, I., and Hinton, G. ImageNet classification with deep convolutional neural networks. In *Proceedings in Advances in Neural Information Processing Systems*, Lake Tahoe, NV, 25: 1090–1098, 2012.

27. Simonyan, K., and Zisserman, A., Very deep convolutional networks for large-scale image recognition. *arXiv preprint arXiv:1409.1556*, 2014.

28. Chollet, F., and Allaire, J., *Deep Learning with R*, Manning Publications Co.: Greenwich, CT, 2018.

29. Hemanth, D.J., and Vieira Estrela, V., *Deep Learning for Image Processing Applications* (Advances in Parallel Computing), IOS Press, 2017, December 31.

30. Raaijmakers, Stephan, *Deep Learning for Natural Language Processing*, Manning Publications, Shelter Island, NY, 2020, May 12.

31. Kamath, Uday, Liu, John, and Whitaker, James, *Deep Learning for NLP and Speech Recognition*, Springer, New York, NY; 1st ed., 2019, June 24.

32. Sadler, Matthew, and Regan, Natasha, Game Changer: AlphaZero's Groundbreaking Chess Strategies and the Promise of AI, *New in Chess*, Alkmaar, Netherlands, January 25, 2019.

33. McGrath, Michael E. *Autonomous Vehicles: Opportunities, Strategies and Disruptions: Updated and Expanded* Second Edition, Independently published, 2019, December 2.

34. Eliot, Lance., AI self-driving cars divulgement: Practical advances In *Artificial Intelligence And Machine Learning*, LBE Press Publishing, 2020, July 9.

35. Goodfellow, Ian, Bengio, Yoshua, and Courville, Aaron, D*eep Learning (Adaptive Computation and Machine Learning Series)*, The MIT Press, Cambridge, MA, 2016, November 18.

36. Ibrahim, D., An overview of soft computing. *Procedia Computer Science*, 102: 34–38, 2016.

37. Iba, Hitoshi, *Evolutionary Approach to Machine Learning and Deep Neural Networks: Neuro-Evolution and Gene Regulatory Networks*, Springer, New York, NY, 1st ed., 2018, June 26.
38. Tad, Gonsalves, *Artificial Intelligence: A Non-Technical Introduction*, Sophia University Press, Tokyo, 2017.
39. Covington, Paul, Adams, Jay, and Sargin, Emre, Paul Covington Jay Adams Emre Sargin, DNN for YouTube Recommendations. In *Proceedings of the 10th ACM Conference on Recommender Systems*, ACM, New York, NY, 2016.
40. https://analyticsindiamag.com/10-open-source-datasets-one-must-know-to-build-recommender-systems/.

Chapter 6

Intelligent business processes with embedded analytics

INTRODUCTION

Businesses are made up of business processes. These business processes are a series of activities with inputs, outputs, roles, and goals. Business processes provide value externally (e.g., customers) and internally (e.g., staff). Business optimization (BO) aims to achieve the process goals efficiently and effectively by embedding artificial intelligence (AI). BO is a business initiative that starts with the evaluation of the desired business outcomes. These outcomes are described in the strategy and planning effort (discussed in Chapter 3). The evaluation of business outcomes is followed by the development of a set of process maps or models. These process models help in understanding where and how to embed AI; the models also assist change management. A model-based, performance-driven approach eases the understanding and application of handset, dataset, toolset, and mindset (discussed in Chapter 2) to business processes.

Processes are modeled keeping the goal of the user in mind. Serving a user efficiently, providing an enhanced user experience, and having an efficient and optimized supply chain are examples of user value. AI analyses data and provide timely and accurate insights for optimized decision-making. In turn, these business processes also provide value for the organization.

A process is the "manner in which" a suite of activities and tasks are carried out. Processes are thus the "how" aspect of an organization's functions. The manner in which a bank account is opened, a patient is admitted to a hospital, or a sales and marketing campaign is undertaken are examples of processes. Business processes can become highly complex and intertwined with processes from other collaborating organizations. Processes use technologies and people to achieve their goals. Embedding AI in business processes makes them efficient and effective.

Intelligent business processes are flexible as they adjust to the needs of their users quickly and interactively. This flexibility of a process provides for business agility. Business agility provides accuracy, speed, and distributed decision points.[1] Agility is enhanced by the speed and accuracy of AI-based analytics to provide insights.

Intelligence is also a factor of sensitivity of analytics. Insights are needed within a given time beyond which they lose their agile value. This sensitivity in analytics is AI's ability to analyze increasingly finer granular data in shorter time spans. Finer, granular analytics enhance the ability of decision-makers to respond quickly and effectively to stimuli. Agility is also the ability of these analytics to experiment with variables in real time. An agile approach to embedding AI in business processes enables experimenting, showcasing, accepting feedback, and modifying the analytics to suit the changing user needs. Agile principles and practices together with analytics help the organization respond rapidly and accurately. Agile as a culture anticipates issues before they arise. This is because of trust, visibility, and iterations in an Agile culture. The explicit (i.e., objective) and tacit (i.e., subjective) aspects of decision-making are judiciously combined in intelligent business processes.

Enterprise systems encompass subsystems, large amounts of heterogeneous data, information, people, technologies, diverse applications, and processes. AI is applied to analyze Big Data in such enterprise systems.[2] Big Data mining is similar to data mining but the scale of Big Data is much larger.[3] Challenges in the use of Big Data include storing and analyzing large, rapidly growing data; handling diverse data sources and stores; and deciding how to utilize that data.[4]

Excellence in business process modeling includes their unification. This unification results in ease of use for the customer as a single point of contact is created. Figure 6.1[5] shows a comparison between (a) knowledge workers and customers using multiple silos of data and (b) empowered knowledge workers and customers through self-serve analytics. Embedding Big Data analytics in business processes enables business agility.

Enhanced knowledge sharing of insights occurs with formal process models. Insights are generated and recorded within the business processes which, in turn, are shared by other users. Process optimization reengineers processes that change the way the organization operated. For example, the cash withdrawal process uses digital technology to eliminate the printing of a physical receipt. Analytics can reduce the time spent by a customer in the queue for physical cash withdrawals by applying the algorithms to predict the size and type of withdrawal transactions. Data analytics provide rich insights that initiate and enable these changes.

Intelligent business processes also support a Lean-Agile organizational structure in which the organization has a flattened hierarchy and decision-making happens at the place of action. As a result, individuals at all levels of the organization are empowered to take decisions based on their responsibilities. This localized and decentralized decision-making is a major change in the style of working resulting from AI.

Appropriate level of insights and creation of audit trails frees the knowledge workers to make decisions. Intelligent processes support decision-making at

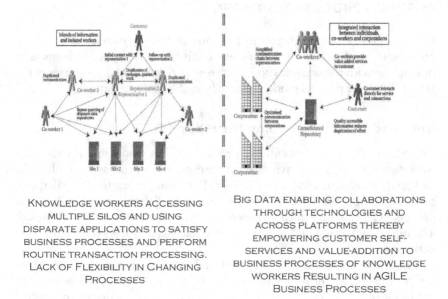

KNOWLEDGE WORKERS ACCESSING MULTIPLE SILOS AND USING DISPARATE APPLICATIONS TO SATISFY BUSINESS PROCESSES AND PERFORM ROUTINE TRANSACTION PROCESSING. LACK OF FLEXIBILITY IN CHANGING PROCESSES

BIG DATA ENABLING COLLABORATIONS THROUGH TECHNOLOGIES AND ACROSS PLATFORMS THEREBY EMPOWERING CUSTOMER SELF-SERVICES AND VALUE-ADDITION TO BUSINESS PROCESSES OF KNOWLEDGE WORKERS RESULTING IN AGILE BUSINESS PROCESSES

Figure 6.1 Embedding AI in business processes resulting in agile business processes.

the lowermost rung of the organization. As both internal and external business processes become optimized, a sustainable organization with reduced environmental footprint emerges. Process models provide not only cost and time benefits to the organization but also advantages in terms of environment and sustainability.

Business analysis is the modeling of these processes to identify opportunities to embed data analytics for optimization. BO is a fundamental rethinking and radical redesign of business processes[6] to achieve dramatic advantages of business agility. Organization-wide, holistic, and dynamic approach to optimization provides these dramatic business advantages. Optimization of business processes requires the creation of process models using standards such as the BABOK of the IIBA.[7] Process models are created by business analysis (as against analytics) in order to examine the activities and tasks that can be (a) made redundant due to analytics and (b) enhanced in terms of their accuracy and speed. Business analysis deals with changes to *the way* in which business is carried out.

Each business function and its processes are investigated by business analysts for opportunities to embed data analytics. Process transformation and change management follow these investigations. These changes include embedding AI-based methods within business processes. Examples of these methods as discussed by Dhar and Stein[8] are use of genetic algorithms, neural networks, expert reasoning, fuzzy logic, and case-based reasoning.

BUSINESS PROCESS MODELING

Business processes are integral to BO. Business is organized around these processes. A value-generating business focuses on the users, their goals in interacting with the organization, and their perception of a good user experience. Business process models enable business optimization.

Business process modeling (BPM) in BO

Embedding data analytics within a process requires modeling its goals, roles (stakeholders), inputs, activities, and outputs. Each process model has the end-goal of providing value to the user. For example, customer, teller, and branch manager form the starting point for modeling a banking business process. The end goal of this process is, say, a successful cash withdrawal. The inputs, activities, and outputs are modeled as a process flowchart or an activity diagram.[9]

Business processes can be external and internal to the business. Some business processes form the core, others are the supporting processes of an organization. External facing business processes provide value to an external stakeholder, such as a customer in a bank or a passenger in an airline. Use cases[10] for external-facing processes include "Customer withdraws Cash," "Passenger Checks-In a flight," and "Patient admits to a hospital." Internal business processes satisfy the needs of an internal stakeholder, such as a branch manager in a bank or a scheduler in an airline. Use cases for internal-facing processes include: "End-of-Day Cash-in and Cash-out," "from the branch," "Scheduling of flights," and "Re-order antibiotics."

Internal business processes supporting the business include operations and management (OSS), production (parts of ERP), HR and people (HRM), legal and compliance (GRC), and Carbon Emission Management Systems (CEMS).

External business processes are facing the customers and suppliers and deal primarily with the growth of the organization. Typical examples include CRM, SCM, ERP, and BSS. The entire suite of processes is studied and modeled as part of BO.

The analytics within these processes can remain at a higher macro level or, alternatively, drill down to a fine granularity. Coarse, granular analytics work on large volume, relative static historical data. This is called descriptive analytics which provides a historical understanding of process performance. Fine granular analytics provide insights for micro actions and snap-decisions with pinpoint accuracy – and are usually predictive in nature. Fine granular analytics need current data and the present business context in which decisions are made.

Analytics embedded within properly modeled business processes is the key to BO. Process models help in deciding which type of analytics (descriptive, predictive, or prescriptive) will provide the highest value to the user. Process models also assist in determining the level of granularity. Optimization of business processes is an ongoing activity that improves performance with each iteration.

Change management processes

Processes also enable, aid, and support the transition of an organization to AI-enabled optimized business. These are change management processes that are not an internal or external business process. Instead, they are "supporting" processes that help the organization with its BO effort. These adoption processes impact, change, and transform the internal and external business processes. BABOK 3.0[11] contains a detailed description of "The Transition Requirements" for the business processes and associated documentation.

The organizational structure (hierarchy) also plays a part in the agility of a process. The primary stakeholder in a process is involved in "obtaining" value, and stakeholders from within the organization are involved in "providing" value. In a data-driven business, processes are executed based on insights from data analytics. Each process is also extensively supported by technologies, systems, IoT devices, and security and privacy. These aforementioned support elements need to collaborate in order to optimize processes.

Successful collaboration requires the process modeler to pay attention to the stakeholders, their activities, the dependencies between activities, and the end deliverables. Involvement of these key external stakeholders earlier in the adoption helps undertake collaborative arrangements (and agreements) for business processes executed across multiple organizations. The end result is a suite of lean and agile business processes with embedded analytics. Supporting processes enable further sharing of problems, solutions, and knowledge across the organization. Finally, training the staff to use the process is a mindset requirement.

Composite agile method and strategy (CAMS)

The techniques (practices) of Agile play an important role in modeling the new digital business processes with embedded analytics. Composite Agile Method and Strategy (CAMS)[12] combines Agile techniques together with the formal, planned methods for process modeling. Composite Agile Method and Strategy (CAMS) in BO provides the benefits of visibility and transparency, change management, integration of solution, and quality through continuous testing and showcasing.

This use of composite Agile enables iterative and incremental approach to process modeling. The Agile iterative approach also helps an organization build its technical and analytical skills and capabilities for BO. Iterations and showcasing of process models (following Agile) also involve collaborating business partners and supporting technology vendors. These iterations help reduce the risks in BO. Agile as a method also provides an understanding of the capability and maturity of the organization. Changes are continuously showcased to the users. Initial experimentation, alignment of analytics with business processes, and alignment of technologies with Enterprise Architecture (EA)[13] reduce the risks in BO.

Process maturity of an organization is another important factor in reducing risks. This maturity of the organization undertaking BO is ascertained by examining the technical capabilities, economic strength, people skills and attitude, and the business processes of the organization.

BUSINESS PROCESS AGILITY

Agile is important in the discussion on business processes and their optimization. Optimized business processes are agile. Agile started an iterative and incremental approach to developing software solutions. Agile has evolved into a business characteristic.

Lean-agile processes

An optimized business is both lean and agile. Analytics enable these business characteristics to flourish. Lean approaches are also used for process optimization, outsourcing strategies, and greening an organization[14] – and their impacts studied on enterprises, government, and society. Process reengineering considers "Lean" as a means to reducing and/or eliminating wastages within the organization processes. Large and global organizations benefit immediately through "Lean" as they tend to have substantial wastages within their inventory and supply chains. For example, large-scale processes and their corresponding value-streams (such as those from mining, agriculture, and airlines sector) can be reengineered by introducing AI-based analytics in them, applying metrics and measurements, and training their users. Agile as a culture also helps the business processes of an organization to become Lean.[15] This leanness of business processes is the result of analytics embedded within the processes. This is so because analytics improve accuracy, reduce time, and decentralize decision-making. As a result, many redundant activities are identified and removed from a business process.

Visibility and transparency

While Agile can help produce faster and higher-quality solutions, these solutions need to be made visible to the end user *while they are getting developed* and not at the end of their development. Agile makes the analytical models and the corresponding business processes immediately visible. The

Kanban[16] boards in agile enable the users to see progress of the development and provide immediate feedback. This visibility also provides transparency of the progress of BO to all stakeholders.

Change management

Agile methods welcome change rather than resist it. This attitude inculcated by Agility is most helpful in transforming business processes. This is because the change in the business processes due to embedding AI in them is incremental and transparent to all stakeholders.

Integration solutions

AI-enabled solutions can become complex. These solutions also need collaboration. Each solution is dealing with multiple other solutions, platforms, infrastructures, devices, and networks. Agility in developing solutions enables the inclusion of external interfaces (API) as a prototype, thereby facilitating and enhancing collaboration and integration. Agile as a culture also opens the doors for physical collaborations with partnering organizations. This results in innovative and collaborative business processes.

Quality through continuous testing and showcasing

Testing includes functional accuracy and nonfunctional or operational quality (e.g., performance, security, and usability). Quality of business processes needs verification and validation as they change through embedded analytics. Such comprehensive testing and continuous showcasing are facilitated by Agility in developing AI solutions. Testing the business processes in an "operational" environment is vital for satisfying user experience, and that happens incrementally through Agile.

DATA ANALYTICS AND BUSINESS AGILITY

Agility is the flexibility within the organization's processes. Agile business changes rapidly in response to changing business situations. Data analytics, embedded with the business processes, are able to provide both descriptive and predictive insights to the decision-makers. As a result, businesses can anticipate challenges and opportunities and improve their responses accordingly. Lean and Agile are business characteristics that are enhanced by the application of analytics to the business processes. Business processes are supported by analytics, technology, infrastructure or applications, and skilled users.

Embedding analytics within business processes changes the shape of the process, allows a business to focus on its core offerings, and "off loads" noncore processes to supporting partners. This offloading results in lean-ness of business processes, which also makes them agile.[17]

Decentralized decision-making

Embedding analytics in decision-making at all levels of a business decen-tralizes the decision-making process. This makes a business more respon-sive as many decisions, guided by analytics can be taken immediately. Organizations are better able to adapt to changes in their environment due to decentralization.

Finer granularity in business response

Fine granular analytics improve agility and accuracy of a business response. The finer the granularity, the more personalized is the response. This is so because finer granularity provides pinpoint accuracy for decision-making.

Elimination of redundancies

Removal of business processes that have otherwise become entrenched within the organization and may be redundant occurs due to formal model-ing and examination of those business processes. Activities and processes that do not add value to their users and which are duplicated elsewhere in the organization can be removed as part of reengineering of the organization.[18]

Enhancing sustainability in operations

Analytics are instrumental in making the processes lean. This is so because, with the help of analytics, processes eliminate wastage and redundancy. This reduction in wastage and improved monitoring and control of the pro-cess also reduce the carbon footprint of the organization. AI plays a role in improving the sustainability of the organization.[19]

Risks, compliance and audit requirements

Governance-risk-compliance (GRC), quality, and audit support processes ensure that Big Data analytics embedded in the internal and external busi-ness processes are compliant with the many legal and privacy requirements, are able to manage security and privacy of the data within them, and are subject to traceability and audit.

Complex regulatory requirements include security, compliance, audit, and risk management – and they consume substantial business resources. An orga-nization has to specifically consider and satisfy these requirements. CAMS makes provision for a detailed documentation to satisfy these requirements.

Disaster recovery (DR)

Data analytics enable predictions of disasters based on (a) past historical data of an organization, (b) external third party from the industry, and (c) alternative data sourced from vendors. DR also includes elements of prescriptive analytics as it recommends actions to be taken in the event of a disaster.

BUSINESS ANALYSIS & REQUIREMENTS MODELING

Business analysis (BA) is a discipline in its own right. The IIBA[20] provides a framework for BA work that extends beyond requirements modeling (RM). Creating process models of the business processes is important in the BO exercise. This is because BO is business oriented and not limited to technologies of AI. A few important BA techniques that help BO include critical thinking, the art of questioning, and mind mapping are described here.

Critical thinking in BPM

Critical thinking[21] is undertaken to arrive at the root cause of an issue and is a core activity of business analysis. Challenging current thinking and beliefs to determine what is true, partially true, or false is important during process modeling. For example, a business analyst thinking critically will question the purpose of a process, the relevance to the stakeholders, and the analytics to be used. Critical thinking underlies the other business analyst problem-solving thinking modes.[22]

System thinking accompanies critical thinking as it views problems as a whole system. It is holistic and, therefore, more strategic. Through systems thinking, the business analyst views the organization and its business processes in its entirety to truly understand the impacts of the use of AI-based analytics and of the changes to the organization.

Critical and system thinking enable BA to handle the long-term impacts of AI adoption.

Art of questioning

Business analysis starts by asking the right questions related to a business process. Questions are asked in the right way and at the right time to the right people.[23,24] Questions typically include What? Why? When? Where? How? Who? BA's base their questions on clarity, precision, accuracy, relevance, and depth. Such questions help understand ambiguity and complexity within the business processes. Business analysts ask questions, listen to the answers, document and model them, and use the results for appropriate actions.

The purpose or goal in asking questions can be divided roughly into two categories:

Understanding the Problem – to ascertain the business needs, capabilities, requirements, and expected levels of quality in a business process. Needs analysis, root cause analysis, SWOT[25] and PESTLE[26] analysis, and critical thinking are examples of techniques used in exploring a business problem. These techniques benefit from a suite of questions that are themselves exploratory and subjective in nature. Techniques in requirements modeling, such as user stories and use cases, also require the business analyst to employ the art of questioning to arrive at functional models of the requirement.

Providing the Solution – the purpose of these questions is to arrive at an appropriate solution to the problem explored. Solutions encompass architecture, design, developing, testing, and deployment in a typical software project. Examples of standards used in a software solution production include UML,[27] agile (methods), TOGAF,[28] development environments test standards (e.g., ISTQB),[29] and deployment approaches. Solution space questions are far more objective and narrower than in problem exploration as they focus on a particular technology used in the solution.

Machine learning to frame questions

The art of questioning undergoes some modifications in the AI model. Traditionally, businesses have a rough idea of what the problems are and what they do not know. With Big Data, however, both the problems and the possible hidden answers in the data mass are unknown. For example, business leaders can ask the data scientists (or equivalent roles) whether it is worth exploring customer demographics to figure out customer attrition or potential growth or another potential topic. Can the data itself give an indication of what can be asked of it? This is precisely where machine learning (ML) has a role to play.

The size and complexity of Big Data is such that building the models of solutions from Big Data is not enough; tools are needed to handle the answers and also frame new questions. With Big Data, business decision-makers struggle to figure out what questions to ask. With the exponential growth in data, it becomes important to not only provide better answers to business questions, but also help businesses understand what types of questions they must ask of their data.

ML algorithms can dive deeper into data in order to identify patterns that are not possible to discern with traditional analytical approaches. This is because of the multilayered or tiered nature of hidden, vast amounts of data. Algorithms need to be created in a manner that creates learning through the execution and provides insights. Thus, ML is the dynamic use of algorithms, which provides a range of opportunities ranging from learned user behavior to dynamic cyber defense.

ML algorithms search through massive amounts of data to find relevant and interesting questions that the collaborative business can ask to improve its overall value proposition. Such crucial questions are the result of identifying insightful similarities and differences within a large, multidimensional dataset that is related or connected with each other in a collaborative enterprise system (CES).[30]

Mind mapping

A mind map is an important BA tool. Mind map is used in BO to show the relationship between data, concept, user, and the process. Mind map is an excellent tool for visual thinkers who are trying to establish the business value of data. Business analysts think, analyze, model, and present processes and their elements in images and pictures rather than words. Business analysts use the graphical representations of mind maps to visualize the problem and the corresponding solution. Mind map provides the opportunity for communication very well.

A mind map starts with the central idea of a data item, a concept, or a business process. Concepts are added as extensions to the central concept developed. Process models and requirements documents are better organized in BO by drawing the mind map. This is so because mind maps shift the focus of BO to business and away from AI technologies. Figure 6.2 shows the example of a mind map. This particular example starts with a central theme: currency options predictive analytics. Such predictive analytics needs

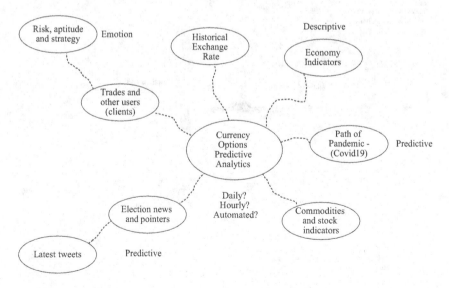

Figure 6.2 Mind map of a data-driven business process in financial market prediction.

data such as historical exchange rates, the economic indicators, path of pandemic, commodities and stock indicators, election news and pointers, trades and other users and risk, aptitude, and strategy. Note how the central theme or concept is expanded based on an extension of the previous node. This expansion of a mind map is also iterative and incremental, following the Agile approach to developing BO solutions.

Comparison of processes for gaps

Use cases, activity diagrams, and mind maps are the tools to create a model of the business processes. These process models can be created for the current and future processes. The future processes are the ones that are embedded with AI-based analytics. The gaps are identified by comparing process models. Metrics are used in BO in order to (a) justify the initiative to embed data-driven decision-making in the business, (b) prove the success of the initiative, and (c) reduce risks. Metrics also help in comparing process models. Figure 6.3 shows an example of a basic comparison of a patient check-in process with and without AI. Process models help in identifying gaps in the value proposition of the process.

Process models keep the focus of the BO initiative on customer value. Treating AI implementation as a technology project is a risk. This risk is ameliorated by use of the business analysis (as against analytics) capabilities in BO.

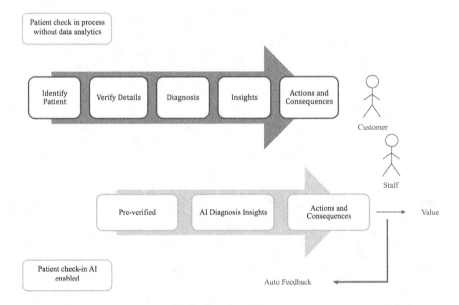

Figure 6.3 Comparison of current process with a digitally enabled smart process.

Managing business system changes

While the business processes, represented by the use cases, activity diagrams, and mind maps, change with embedded analytics, the various systems to support the business functions have to absorb newer and finer granular analytics that support changes to the business processes. Furthermore, collaborative business processes expose an organization's systems and applications to those of the collaborating partners. This embedding of analytics and extension of electronic collaboration changes the systems and their interfaces, and requires additional focus on the cybersecurity aspects of the systems.

Each analytic embedded in the above processes leads to further levels of analytics. Examples of this further analytical work include enabling customer segmentation and understanding their attrition rates, working out the formation of communities in order to improve product acceptance and use, running targeted marketing campaigns through multiple channels, developing accurate strategies for cross- and up-selling, and optimizing business processes.

Following business functions and systems undergo a change due to business optimization using AI and data analytics:

- Customer relationship management (CRM) – Changes the way products are priced based on a detailed understanding of customer buying patterns and profiles. Customer analytics search for data based on a customer footprint on social media and mobile to improve customer understanding. This understanding includes the customer demographics, thereby improving the organization's ability to provide such customers with a holistic view of their own requirements and enabling the organization to strategize for the satisfaction of its customer base.
- Marketing and promotions are based on fine granular analytics and targeted at specific customers. Data analytics provide a strategy to market a product to a fine cross section of customers and then replicating it over a large cross section of other customers. Learnability of analytical solutions is an important aspect of agility in marketing. Small, experimental introduction of marketing effort, receiving instantaneous feedback, and incorporating that feedback in subsequent iteration of marketing effort is the way this analytics is leveraged for business agility. Customer sentiments are based on Net Promoter Score (NPS)[31] and can help ascertain what the customers think and feel about the product. As a result, products and services can be changed quickly to accommodate the changing sentiments. Assessing customer sentiments using AI and ML automated tools and presenting appropriate products to them are major changes.
- Analyzing prices indicates how to prioritize and sell products based on discriminatory pricing. The creation of such fine, discriminatory pricing that can change from moment to moment and from one individual

to another is a vital element of business agility. Since customers can also undertake self-service analytics, the face-to-face contact with the business is further reduced.

- Supply chain management (SCM) – AI helps in optimizing the inventories and understanding the logistics and deliveries. This optimization changes the way the SCM works. Supply chain processes are based on data associated with quality, fault reports, and also product recalls should a product fail a compliance need. These analytics can also assist in process automation as they enable changes to the product directions in an automated or semiautomated manner depending on the feedback being received. This optimization also reduces wastage due to poor inventory management. Collaboration and partnership with suppliers are established through negotiations and contracts as a result of use of analytics electronic implementation of these political speeds the SCM processes.

- Enterprise resource planning (ERP) – AI-based analytics improve the understanding, creation, launching, management, and withdrawal of products and services. Insights generated by Big Data analytics help in optimizing these ERP functions by ensuring their timelines and accuracy. Products are added or removed in an agile, iterative manner with the help of analysis.

- Sustainability applications – usually under the umbrella of Carbon Emission Management System (CEMS) – These functions help handle corporate responsibility and sustainability requirement by accurately calculating and controlling the carbon footprint of the organization. Data-driven insight enables justification for environmentally conscious decisions by the business. Data is able to show to the decision-makers the carbon footprint of each business activity and the organization. Data analytics also correlates the carbon footprints to the risks of fines due to noncompliance.

- Billing support system (BSS) – Billing, finance, and accounting systems change with Big Data as the analytics are able to spot trends and advise the accounting functions in terms of anticipated changes, trends in billing errors, and where to focus the resources to improve billing efficiencies. These changes have a direct impact on delivering customer value.

- Operations and support system (OSS) – Applications, networks, databases, scheduling, security, and other operational support provided by the systems are analyzed with the help of data – typically collected through IoT sensors. Data analytics provide a formal, data-driven approach to maintenance and upkeep of production equipment. This can be an important type of analytics with applications in aircraft maintenance, bridge repairs, and large residential and industrial structures.

- HR and people (HRM) – Organizational structure, motivation, morale, rewards, KPIs, and collaboration and partnership with external organizations are all affected by the use of data changes. Recruitment itself gets shifted to a sophisticated analytical filter that goes way beyond the keyword searches.

- Governance, risk, and control (GRC) – Through accurate analytics that enables the organization to comply with regulation (such as with SOX legislation). Risks are identified much earlier than they occur due to predictive analytics. Systems are also able to imitate autorecovery functions on some processes in the organization.

EMBEDDING ANALYTICS IN BUSINESS PROCESSES

Embedding analytics in business processes requires (a) modeling of processes and (b) understanding of the analytics. The range and types of analytics and corresponding supporting technologies create opportunities for BO. Data analytics are traditionally divided into descriptive, predictive, and prescriptive analytics. These analytical categories are based on their purpose and use in business. For example, predictive analytics plots a trend and attempts to show the position of a point in time (future). This type of analytics would require assembling large historical data, integrating it with current data, developing experimental analytical models, and incorporating the feedback from showcasing the results. As a result, this type of analytics may not be ideal for situations requiring real-time, fine-granular insights. Alternatively, descriptive analytics will be more ideal for the aforementioned scenario.

Preparing the data

Most practical analytics are performed on both structured applications (e.g., existing CRM, ERP) and unstructured information sources from Web pages, blogs, and contact centers. Social media and mobile (SoMo) are rich sources of alternative data. Such unstructured data requires preparation before it can be analyzed. MapReduce, "R," Python, and similar tools process unstructured information, add structure to it, and set a stage for its analysis. The extent to which the unstructured data needs to be "prepared" for analytics is an important consideration in BO. This section discusses categories of analytics in order to identify their appropriateness in practice. These categorizations also indicate how the analytics will benefit the business. The same analytical tool or application can be used in multiple analytical categories.

Data analytic types and relevance in BO

Analytical categories help in understanding how they can be applied in practical BO. Analytics helps an organization become and remain Agile.

Analytics themselves help the business by embedding agility with its business processes and organizational structure. The Agile methods also help in the solutions space in developing the analytics. Agility for the business is thus supported by developing analytics with an agile approach.

This list of analytics (Table 6.1) helps understand the possibilities of richness in analytics as well as their application in optimization. For example, if the purpose of an analytics is to describe a past (historical) situation, then it's an informative-descriptive type analytics. Alternatively, if a trend plot is used to anticipate a change in weather pattern next week, it is a predictive (futuristic) analytic. Table 6.1 summarizes these analytical categories and corresponding business optimization strategies.

Each of the analytics summarized in Table 6.1 incrementally builds on the previous type of analytics. The following is an additional description of the three important categories of analytics and how they provide value in business optimization.

Table 6.1 Data analytics types and corresponding strategies for BO

Analytics (Type)	What it implies	Strategies for business process optimization
Informative	Available as easily (publicly) available information that can be broadcast.	Being static output, has minimal impact on processes. Speed of processing on a large data set is the minimal cybersecurity requirement.
Descriptive	Describe and Broadcast Information/insights based on past, historical large data analysis. Requires preparation of data that may be in different unstructured format. May take a long processing time.	Provides information to a large group of customers depending on their need. Customer contact points are increased. Business users can better understand the situations in their business processes.
Diagnostic	Define analysis as highly parameterized and focus on a well-defined function to diagnose errors, gaps, and failures.	Enables identification of preemptive problems. For example, equipment failures in a manufacturing unit.
Predictive	Analyze the data based on an event and provide predictions to enable immediate response. Key Stakeholder focus in order to help their decisions. Helps in risk management as many risks are not visible.	Provide decision-makers with trends and patterns of the future. Time and accuracy sensitive. Depends on the granularity of business processes. Innovative risk reduction by using continuous, iterative analytics to explore options; multiple "what-if" scenario generation using data.
Prescriptive	Decision-maker combining with tacit intelligence.	Decentralize decision-making based on recommendations combined with experience.

Descriptive analytics

Descriptive analytics examines large amounts of historical data to describe the current scenarios. This is a relatively static data. For example, the describing process of sales and inventories channels for a salesperson to understand the areas to focus on. A description of the performance of an individual salesperson or team is also part of this analysis. Description of current and past performance can be analyzed to determine the reasons for success or failure. Data can be classified depending on similar characteristics (e.g., targeted sales campaign). Examples of visualizations in descriptive analytics include scatter graphs and bubble charts.

Predictive analytics

Predictive analytics is meant to identify and determine futuristic trends such as sales, expected sales, or changes to market behavior. Predictions can be used in capacity planning and customer retention strategies. Statistical models used in forecasting are part of predictive analytics. The future of customer behavior is predicted by examining vast data sets on social media and mobile data to ascertain customer sentiment. Predictive analytics combine unstructured data (e.g., Web logs, blogs, Facebook, and Twitter feeds), together with structured transactional data. Predictive analytics are more fine granular than descriptive analytics.

Prescriptive analytics

Prescriptive analytics goes beyond predications and into the realms of suggested decisions. Decisions using prescriptive analytics are optimized based on the discovery of trends and patterns within the data. Business rules are understood and coded in prescribing actions based on these analytics. Prescriptive analytics make use of past data and models together with the most current data to create a system that can be immediately applied and reevaluated across numerous instances. Thus, prescriptive analytics have a role to play in process automation.

COLLABORATIVE DIGITAL BUSINESS PROCESSES

Digital technologies enable businesses to connect, exchange data and information, execute contracts, and increase the overall value to their customers. This connectivity is usually based on "services." The tools and technologies of electronic collaboration support globally dispersed workforces within the same organization. Collaboration focuses on integrating and extending corporate operations and services beyond a single enterprise.

BO is an ongoing and, perhaps, an unending activity. Collaboration of business processes is a major opportunity for BO. BO is "agile" because it is continuously changing. Business agility is a customer-centric effort that benefits by collaborating with other enterprises in their space to enhance the user experience.

The EA's mechanism to measure, compare, and contrast technology investments provides the technical foundation for the organization. Such comparison also provides strategic input into the comparison of processes, justification of BO, and opportunities for collaboration.

Collaborations require frameworks, patterns, and government and industry regulations as well as best practices, corporate policies, and previously adopted standards. These strategic choices make collaborative enterprises agile and flexible enough to adapt to and manage changes in the corporate IT environment. Ease of integration and the portability of key elements in existing disparate platforms of the partnering organizations are important.

Teamwork as a part of collaboration is necessary to successfully integrate enterprises. Hence, evolving business requirements and corporate strategies together make a strong business case for developing collaborative enterprise architectures.

Collaboration advantage in a digital world

The impetus for deploying a collaborative enterprise is manifold.

First, it reestablishes a basis on which to do business with partners and customers in a distributed world, enforcing a set of ground rules such as service-level agreements (SLAs) or trading-partner agreements (TPAs).

Second, it gives the enterprise architecture and infrastructure the flexibility to handle growing business needs and to connect globally with prospective partners, customers, or vendors.

Third, the collaborative enterprise also lays the foundation for establishing communities of practice and promotes reuse of the available knowledge base without reinventing the wheel.

Business collaborations participate in a mutually beneficial venture. Connecting, communicating, coordinating, and making commitments between two or more businesses widen the offerings range. Depending on established roles and responsibilities in an organization agreements such as SLAs, TPAs, subcontracting, or sourcing agreements are electronically implemented. Collaborations share appropriate information, technology, and techniques using a common electronic medium. Decision-making data and transaction processing information are exchanged in e-collaborations.

A collaborative environment establishes this operational conduit to enable optimization of product and service offerings. A collaborative environment provides common services, a set of information exchanges, and a delivery mechanism. Knowledge management (KM), content management, user experience and relationship management, and security are all interconnected through interfaces. Data analytics offer greater changes of accurate and timely insights if they are conducted over large and wide-ranging data sets. This is the biggest advantage offered by collaborators in BO.

Collaborative digital business

A collaborative business enterprise involves people interacting with one another using processes, policies, best practices, and a collaborative environment. Technologies and techniques across the enterprise facilitate collaboration. The human component of the collaborative enterprise emphasizes the complexity of interactions due to the cultural diversity across the organization. Collaborative processes include human-to-system, system-to-system, or system-to-human forms. Collaborative enterprise systems include integration of business applications, business processes, frameworks, and reference models inside or beyond the boundaries of a specific enterprise to provide appropriate access to the participants.

Digital business uses electronic medium to collaborate electronically with customers, partners, staff, and regulators. Therefore, every digital business is essentially a collaborative digital business. Collaborative technologies enable a digital business enterprise to explore many opportunities to provide additional customer value. Collaborative digital business enterprise becomes a complex and challenging entity because its boundaries are fuzzy and they are continuously changing. "Collaborative EA" enables handling the complexity and dynamicity of collaborative digital business.[32] In addition to business-to-business collaboration, even knowledge-customers and knowledge-workers associate electronically. A collaborative enterprise architecture (EA) also offers major benefits in enabling safe collaboration among customers and workers. Collaborations create and synergize intelligence within and across multiple organizations to produce actionable insights for users or end customers.

When business processes are executed across multiple organizations based on electrically defined policies and contracts, they result in automatic collaboration.

Complexities of collaborative digital business

The complexity of collaborative processes arises not only due to multiplicities of applications, but also by the fact that they are serving dynamically changing business processes. Multiple, collaborating applications that are widely dispersed over physical servers (or in the cloud), many different input/output points over fixed and mobile devices, and dynamically changing data comprising text, audio, video, and graphics – all result in a melee of services that are daunting to decipher for sensible decision-making.

Business processes become complex as they traverse the boundaries of multiple organizations in order to collaborate. Assessment and evaluation of collaborative business processes or workflows, modification to business interactions, and understanding the dependencies processes is an essential part of their modeling. Process management also includes collecting such metrics as user access logs and user fulfillment surveys or tracking the volume of collaborative activities.

Ranging from simple instant messaging and chat sessions to real-time audio- and videoconferencing, from video streaming to e-learning to online communities of practice – the complexity is enormous.

Optimized collaborations

Data, information, process, and knowledge are shared by multiple organizations across their electronic boundaries and result in optimized processes. Analytical insights are made available at the right time and place for the participating organizations. Collaborative Intelligence (CI) aims to handle the challenge of sharing across multiple organizations. Integrated and *intelligent* business and operational processes result from optimized collaborations.

Automated and optimized collaboration among a group of businesses is based on CI. Collaboration leverages information sharing across multiple organizational boundaries and in a dynamic manner.

These electronic collaborations are enabled through tools and technologies (typically Web Services and also, increasingly, Analytics-as-a-Service).[33] CI enhances collective value reduces overall costs. The *intelligence* within the processes is used in decision-making in real life by collaborating organizations.

Table 6.2 documents the key concepts of collaborations in enterprise systems. These concepts form the basis of the mechanisms for embedding intelligence within processes. Intelligence within these systems facilitates collaboration between their processes. Big Data and ML for decision-making result in improvement in collaborative enterprise systems.

Intelligence can be garnered through information technologies that generate new and dynamic knowledge within and across the organization. Collaborative organizations interact with each other, their customers, and suppliers in real-time through web services. In addition to the technical capabilities of software, these collaborations also require strong business relationship-building skills. These business relationships include people skills and forming electronic policies that can be used in creating and executing electronic collaborations. Building relationships and collaboration also leads to closer scrutiny of the inner workings of member companies,[34] resulting in a need for a greater level of trust and mutual understanding between these "clustering" member companies.

E-collaborations happen on a platform that facilitates direct information exchange among otherwise siloes applications (both within and outside of the organization). Internet-based exchanges resulting in sharing of information

Table 6.2 Key concepts of collaborations in enterprise systems[2]

Key concepts	Description
Collaborations	These are essential interactions between businesses (and their systems) in order to carry out business actions and achieve business goals which cannot be achieved by a single business.
Enterprise systems	These are software systems supported by corresponding data that enable business organizations to carry out their key functions (e.g., sales, marketing, inventory management, accounting, HR). Enterprise systems are continuously evolving to incorporate the Cloud, data science, AI, and mobility – together with cybersecurity.
Artificial Intelligence and Machine Learning	Concepts and algorithms that are implemented in order to capitalize on the abilities of software systems to "learn" in both supervised and unsupervised manners and continue to learn and correct themselves.
Big Data	High volume, rapid velocity, and differing varieties of data that has the potential to be analyzed using data science techniques in order to provide insights.
Business decision-making	Includes processes and procedures within business organizations that have the potential for ongoing improvement through the use of intelligence based on (typically) Big Data.

among those applications also need to be facilitated within the Collaborative Enterprise Systems (CES). These information exchanges evolve into ontology-based collaborations among multiple applications and databases.

Furthermore, an organization needs to collaborate among its people and processes. Business value is derived by enabling people to make productive use of applications that go beyond the specific transaction they are engaging in with the organization. Sharing data and information across organizational boundaries can produce imaginative new pieces of knowledge that the organization can creatively use. Collaborative Intelligence (CI) is the extension and application of BI together with collaborative business process engineering,[35] which is built on AI. CES, equipped with ML, can optimize organizational resources by using current cloud computing and SOA capabilities.[36]

VISUALIZATION AND BUSINESS PROCESSES

Visualization is the presentation aspect of BO. Business decisions rely on data analytics. The velocity, volume, and speed of analytics have an impact on the comprehensibility of analytics. Without due consideration to visualization design, making sense of the hidden patterns is impossible no matter how good the analytics are. Presenting the results of analytics in a human-comprehensible way is crucial for the success of BO. How much information is comprehensible? While the metrics vary depending on the user, the classic "span of control"[37] is a good metric to keep in mind. An aesthetically pleasing model will have no more than seven elements. The end users need much less information than the business decision-makers.

Effective visualization is designed in a way to provide the right amount of information, at the right time to the right person, keeping in mind, limitations of the device.

Visualization tools interact with the analytical tools to present the results of the analytics. Ineffective visuals can also be counterproductive as they can reduce the efficiency and effectiveness of decision-making. User-centered designs ensure users are able to learn the interface intuitively and that the interface grows with the growing expertise of the user. Educating the user in the use of visuals adds to its quality.

Workflow models become important in designing group visuals. Visuals are not just analytical results. The movement of a patient in a hospital, for example, is a powerful visual to help the staff.

Presentation of the data results analytics consider three aspects of quality: syntax, semantics, and aesthetics. Syntactically quality deals with factual correctness, whereas semantics deal with the correctness of meaning. Aesthetics implies style and representation of insights from analytics. Aesthetics also represent the ease of creating and reading an analytical model. Although the analytic may be accurate (syntactically) and meaningful (semantically), its useability depends on its visualization or the style of representation.

Designing visualization quality considers the following factors:

- Type and size of devices. Each device has its limitations that will limit the presentation. Visual design includes as many possible devices with varying screen sizes and different operating systems and environments.
- Rate of change of underlying data and corresponding speed of analytics to keep up with the changing data. This will impact the performance of the visualization because the moment the data is updated, the analytics, and therefore the presentation, has to change. Presentation is thus related to the underlying analytics. Flexibility in visuals is enabled through parameters and rules.
- The context dependency of the results being presented in which business scenarios with the results will be used. Thus, the underlying theme or purpose of a visual is crucial. The purpose of a business process has to be reflected by the visual. Thinking and designing the visual starts with the nature and purpose of the business process.
- Amount of information that needs to be shown to the user. This information can change from user to user. Displaying the maximum possible information on one screen is not always a good idea. Aesthetics includes showing relevant and limited information to the user.
- Currency or time duration for which the visualization is relevant. Understanding this currency is critical as it influences the quality of

presentation and also has security implications. If a visual is relevant only for a few seconds, a large, colored, graphic format (rather than numbers and texts) is advisable. Other pieces of information that may be relevant for a longer period of time can be placed in a summarized form on a dashboard to provide both graphics and numbers and text. The time to display and the amount of information to display are both functions of multitiered and multilayered presentations. Workflow between the presentation layers and their interdependencies are also considered in a group (team) visual design.

- Use of sensors and audio cues. Customization of visual presentation is providing users with options to configure. Automation of presentation without configurability option reduces the quality of the user experience. Since each user has her own preference in terms of how she wants to visualize the information, that information should be customizable. Testing of every permutation and combination may not always be possible because of their large numbers.
- Testing – at least initially – can follow the 80–20 rule; that is, 20% of the features and preferences of visuals will be used by 80% of the users. Once those tests are accomplished, the rest of the features can be tested.
- Similar to analytics, visuals also have a level of granularity. Visuals can represent the three middle layers of the data to decision pyramid.

Device and performance consideration in visualization

The end user devices play an important role in visualization. The manner in which visuals are created and displayed depends on the characteristics of the user devices. Following are some important device considerations in designing visualization and presentation of AI analytics to the user:

- Size – Miniaturization has played a great role in the progress of device technology. Miniaturization and consequently portability of devices need due consideration. The art of visualization in analytics has to constantly adjust and rearrange itself to reflect the advances in the size of hand-held devices.
- Security – Security in all aspects of the devices requires due attention. Making the visualization appealing cannot come at the expense of security. Security of devices includes security of the software, application, visual as well as access to the device and its physical security.
- Privacy – Safeguarding user privacy is of paramount importance in visualization. Which visuals to show and for how long should they be shown is a factor of privacy characteristic.
- Functionality – Hardware devices like cell phones and ipads contain large storage and processing capacities. This gives them the ability to

process colored graphics, photos, and videos. The graphic processing power of devices should be fully utilized in visualization.

- Risk – Risk mitigation is another factor in device consideration of visualization. In addition to security and privacy risks, devices also present the risks of inappropriate use and are subject to malware (Chapter 9).
- Speed – As apps get heavier and bulky, their performance on handheld devices can degrade. Slow speed of processing on handheld devices results in one of the most annoying features - the cursor going round and round in circles. Users do not have the patience, hence features such as off-line processing are useful. Apps are also lightweight so that they can be accessed, and their processing interchanged without waiting times.
- Updates to devices – Visuals need to keep pace with the device versions. Older versions of apps should retire and be replaced by newer ones. Users find value in renewal of apps, visualization contents, and interfaces.

CONSOLIDATION WORKSHOP

1. Why should business process modeling and reengineering be considered in the BO initiative? How can process modeling help with AI adoption?
2. What do you understand by the term "granularity"? What is the difference between coarse and fine granular analytics?
3. What is the importance of change management in BO?
4. What is critical thinking? And systems thinking?
5. Discuss the importance of ML in assisting the business with framing the right questions.
6. What is a mind map? Draw a sketch of a mind map for a business scenario of your choice.
7. Why do you think collaboration occurs in a digital business? What are the impacts of collaborative business processes and collaborative decision-making on a business?
8. What challenges exist when embedding analytics in business processes? What are the ways to overcome these challenges?
9. What are the characteristics of the active or dynamic feedback loop in business processes?
10. What are the differences between positioning analytics in the Cloud, on local machines, and on IoT devices?
11. How is automation in collaboration facilitated by service orientation?
12. What are the differences and similarities of automation and optimization of business processes?
13. What is the impact of developing collaborative intelligence as an extension of business intelligence?

14. How does robotics play a role in process optimization (inventory management, etc.)?
15. What are the impacts of digital strategies on the business functions and organizational information systems (including CRM, ERP, Carbon Emission Management System and Governance, Risk & Compliance)?
16. What is BPM and why should it be used in modeling business processes embedded with analytics?
17. How can businesses apply Agile techniques used in Composite Agile Methodology and Strategy (CAMS) and use them within a formal process map for requirements modeling?
18. What should be considered when mining for embedded data in business processes (datafication) and maintaining dynamicity in agile business processes (continuous research)?
19. Why are business processes and decision-making considered "fuzzy"?
20. What is the importance of visualization in effectiveness of analytics?
21. What are the key device considerations in designing visuals?

NOTES

1. Unhelkar, B. 2013. *The Art of Agile Practice: A Composite Approach for Projects and Organizations*, (CRC Press, Taylor & Francis Group /an Auerbach Book, Boca Raton, FL). Authored ISBN 9781439851180, Foreword Steve Blais, USA.
2. Unhelkar, B., and Arntzen, A. 2020. A framework for intelligent collaborative enterprise systems: Concepts, opportunities and challenges. *Special issue of the Scandinavian Journal of Information Systems*, Ed. Eli Hustad, Summer.
3. Lee, H., Chen, K. L., and Yang, J. 2014. The implications of Big Data for the enterprise systems for small businesses. In *The Proceedings of Worldwide Microsoft Dynamics Academic*, Atlanta.
4. Kaisler, S., Armour, F., Espinosa, J. A., and Money, W. 2013. Big Data: Issues and challenges moving forward. In *2013 46th Hawaii International Conference on System Sciences*, Maui, HI, 995–1004.
5. Sherringham, K., and Unhelkar, B. 2020. *Crafting and Shaping Knowledge Worker Services in the Information Economy*, (Springer Nature (Palgrave Macmillan), Singapore), 2019. ISBN 978-981-15-1223-0.
6. Hammer, M., and Champy, J. originally published 1993. Reengineering the corporation, Collins/First Harper Business Essentials, New York, NY.
7. https://www.iiba.org/ accessed 1 Nov 2020.
8. Dhar, V. *Seven Methods for Transforming Corporate Data into Business Intelligence*, 1st ed., Roger Stein ISBN-13: 978-0132820066, (Prentice Hall), New York, NY.
9. Unhelkar, B. 2018. *Software Engineering with UML*, (CRC Press, Taylor & Francis Group/an Auerbach Book, Boca Raton, FL). Authored, Foreword Scott Ambler. ISBN 978-1-138-29743-2.
10. See note 9.
11. BABOK 3.0 from the IIBA – https://www.iiba.org/babok-guide.aspx accessed 6 Dec 2019.

12. See note 1.
13. Hazra, T., and Unhelkar, B. 2020. *Enterprise Architecture for Digital Business: Integrated Transformation Strategies*, (CRC Press, Taylor & Francis Group) (in Production).
14. Unhelkar, B. 2011, April. *Green ICT Strategies & Applications*, (Taylor & Francis, Auerbach Publications, New York/Boca Raton, FL). Authored ISBN: 9781439837801.
15. Unhelkar, B. 2014, April. Lean-Agile Tautology. *Cutter Executive Update*, 3 of 5, 15(5), Agile Product & Project Management Practice, Boston, MA.
16. https://en.wikipedia.org/wiki/Kanban accessed 1 Nov 2020.
17. See note 15.
18. Hammer, M., and Champy, J., originally published 1993, Reengineering the corporation, Collins / First Harper Business Essentials.
19. https://www.iiba.org/
20. https://www.usf.edu/atle/teaching/critical-thinking.aspx accessed 1 Nov 2020.
21. https://www.batimes.com/steve-blais/thinking-like-a-business-analyst.html.
22. Unhelkar, B., and Lyman, C. 2017. Discovering the right questions in Big Data: The Colored de Bruijn graph approach. *Cutter Business Technology Journal*, 30(10/11): 29–34.
23. Unhelkar, B. The art of questioning. *Cutter Executive Update*, Arlington, MA, 15(23), Business Technology Strategies practice (2500 words).
24. https://www.mindtools.com/pages/article/newTMC_05.htm.
25. https://pestleanalysis.com/what-is-pestle-analysis/.
26. See note 9.
27. https://www.opengroup.org/togaf.
28. https://www.istqb.org/.
29. See note 2.
30. https://www.netpromoter.com/know/; https://en.wikipedia.or g/wiki/Net_Promoter.
31. Hazra, T., and Unhelkar, B. 2016, Feb. Leveraging EA to incorporate emerging technology trends for digital transformation. *Cutter IT Journal*, (theme - Disruption and Emergence: What Do They Mean for Enterprise Architecture?) 29(2): 10–16.
32. Yeow, A., Sia, S. K. Soh, C., and Chua, C. 2018. Boundary organization practices for collaboration in enterprise integration. *Information Systems Research*, 29: 149–168.
33. See note 2.
34. Barekat, M. 2001. Virtual-e-teams making e-business-sense. *Manufacturing Engineer*, 80: 66–69.
35. Trivedi, B., and Unhelkar, B. 2009. Semantic integration of environmental web services in an organization. In *2009 Second International Conference on Environmental and Computer Science*, 284–288; Trivedi, B. and Unhelkar, B. 2013. Environmental intelligence and its impact on collaborative business. *BVICAM's International Journal of Information Technology*, ISSN 0973-5658.
36. Gil, D., Ferrández, A., Mora-Mora, H., and Peral, J. 2016. Internet of things: A review of surveys based on context aware intelligent services. *Sensors (Basel, Switzerland)*, 16: 1069.
37. https://www.inc.com/encyclopedia/span-of-control.html accessed 1 Nov 2020.

Chapter 7

Adopting data-driven culture

Leadership and change management for business optimization

LEADERSHIP AND CULTURE CHANGE IN BO

Business optimization (BO) is a strategic management initiative that brings about a fundamental change in the people, processes, technologies, and economies of a business. The organizational structure and dynamics of the business change as it becomes more agile, lean, cohesive, and holistic. The organization becomes more capable of rapid and effective responses to external stimuli and able to initiate its own changes as a result of AI.

A business functions by executing its business processes. This execution of processes provides value to the user. The value can be external or internal to the organization. Optimization, in particular, focuses on the most efficient and effective way to achieve the value. The impact of this AI adoption effort is the changes to the way decisions are made in organizations. For example, with the help of AI-based analytics, customer-facing staff has substantial additional information to make decisions on the spot. Such AI-enabled and reengineered business processes eschew the hierarchical approval processes. AI-based processes flatten the organizational hierarchy and thereby radically change the organizational culture. Managing this change in the cultural mindset of the organization is an interesting challenge in adopting AI, and it mandates astute leadership.

Changing the mindset of people (both staff and customers) is a greater challenge than the reengineering of activities and tasks within a process. Changing to a data-driven culture requires the business leaders to pay due attention to human behavior, motivations, work environment, business operations, and governance. The impact of automation and optimization on the knowledge workers and the customers requires special attention. The manner in which people accept, respond, and operate the AI solutions determines the success (or lack thereof) of the BO initiative. Astute leadership is a crucial element in transitioning to optimized business.

Additional attention is required to the quality-of-service (QOS) factor. Optimized services are affected by quality of data, reliability of devices, and relevance of analytics through to visualization, security, and perception (more discussion in Chapter 8). The customer's perception of quality and security is a supreme indicator of business value. A reliable and resilient business builds capacity and capabilities to handle cybersecurity and quality requirements in an iterative and incremental (ongoing) manner. Business strategies combined with agile instill adaptable behaviors, strengthen skills through training and mentoring, and enable the business to be adaptive, responsive, and resilient.

Leaders are entrusted with envisioning, preparing, and leading the execution of AI-based optimization. Leaders also manage the ensuing change to a data-driven culture. Leaders prepare the organization for risks, disruptions, and unexpected outcomes. Foremost is the people risks that impact staff, users, managers, and support roles such as accounting and audits. This preparedness of the organization before, during, and after business optimization with AI is the key differentiator in providing customer value.

Managing the employees and other contract staff within an organization, keeping up their motivation, and keeping them abreast emanating from the application of AI in optimizing business processes of the changes are crucial ingredients of successful BO. Success in achieving career aspirations by individuals and ensuring personal job satisfaction requires right attitude together with careful planning and subsequent nurturing of that attitude and plan.

The core value of the business dictates what the business is all about. This is the primary reason why the business exists. Leaders ask the question: How will the core value change as a result of BO? Or, does it need to change at all? A radical business transformation can change the core value of the business, for example, from being a nonprofit charity to a profit-making business. Another example of change is that from being an entirely physical business to an online or virtual business.

Leaders create an environment for business optimization that is not limited to technologies. Encouragement and rewards for people, funds, and resources to train and change their mindsets, and adoption of the required frameworks and standards for process improvement, are crucial leadership activities during BO. Culture change in business complements automation and optimization of processes and services.

Change of mindset

Adopting AI-based data-driven culture in an organization requires a change in mindset. Realization of value from AI and real-time decision-making is a result of people implementing and using the solutions. BO requires people to change the way they work, and therefore, their mindset. (This change was discussed in Chapter 2, Table 2.1.) Flexibility of a business is its ability to change internally so as to respond to external pressures. This flexibility is due to the use of AI in business processes. Change is inevitable due to this flexibility. As various systems and functions like marketing, financial management, HR, vendor management, workload distribution, SCM, HR, and compliance undergo a change, it also requires a corresponding change in the mindset. These aforementioned systems change the way they store data and consume the cloud-based analytical services. Advising customers, vendor management, application integration, data feed management, and supplying data to supporting operations require to be managed as they all undergo change.

Changes from automation create less impact than changes from optimization. Routine business processes that are well suited to automation improve the time, accuracy, and number of people employed. Optimization requires greater problem-solving skills and focuses on higher value-added services. Therefore, optimization changes the basic structure of the business process and the value it provides.

Cloud computing, mobile computing, and IoT Big Data analytics and ensuing real-time decision-making provide desired business outcomes so long as knowledge workers[1] with the business processes accept those optimized changes. The right people using the updated processes and making the right decisions at the right time and place require carefully managed change. Developing an approach to the change required of knowledge workers is a leadership responsibility.[2]

Managing the people risk

The complex and changing nature of interactions between people and data-driven processes poses a risk to successful BO. Optimization redefines the roles within the organization including their reporting hierarchy. Roles in an optimized business require advanced problem-solving and complex people management skills that are more challenging than in automation. Changes in business due to optimization require resources skilled in critical thinking, analytical problem-solving, and change management. Users need the capabilities to manage by exceptions and intervene using natural intelligence (NI) in order to manage customer expectations. NI provides the crucial, subjective inputs and also helps alleviate the ethical and moral challenges in data-driven decisions.

Leadership facilitates changes in behavior to reduce people risks. These changes are based on the following considerations:

- Viewing people, processes, and analytics as a holistic combination to provide customer value. People issues to be handled along with AI implementation. This enhances the capabilities of people to use analytics in decision-making.
- Planning for the impact of business and social factors on each other during automation and optimization of processes. In particular, the way the social order changes in the organization.
- The initial hiccups which may result in slowing down of services due to the adoption of AI and managing the expectations of the customers and staff when that happens.

- Providing all necessary training and mentoring to build the capacity and capability for the staff to be adaptive, responsive, and resilient by AI-enabled business processes. Providing coaching to users, including customers, on the new ways of interacting with the organization.
- Developing the organizational ability to anticipate and respond to changes in AI technology by continuously scanning the technology landscape.
- Organizing modeling and implementing business processes based around the users rather than business hierarchy.
- Encouraging modeling of end-to-end process requirements that will provide a clear indication of how people are affected.
- Promoting a holistic quality environment in the organization by using agile iterations and increments.
- Ensuring active leadership participation in the initial planning before commencing AI adoption. Leaders handle the concerns and issues of people even in the planning stages of automation and optimization.
- Leading by example and providing clear vision and direction to the organization. This can be achieved by using AI-based decisions and making them known.
- Adapting formal risk management and a risk-based approach to AI adoption. This includes monitoring risks and preparing to adapt and respond to changes due to automation and optimization on a continuous basis.
- Including metrics, compensation, and rewards in the overall BO strategy.
- Applying Composite Agile in outlining capabilities and delivering projects.
- Supporting the capability of a business to remain resilient and sustain operations through changes due to AI.
- Supporting establishment, management, assessment, and use of collaborative business processes.
- Supporting and providing all necessary training and coaching to the staff.
- Ensuring security and privacy within the AI-enabled business processes. This includes monitoring risks and preparing to adapt and respond to changes due to automation and optimization on a continuous basis.
- Facilitating dynamicity in business processes – change the business processes of the organization to quickly and effectively respond to the changing needs of the customer in a location - and time-independent manner.
- Ensuring corporate responsibility by providing standards and consistency through governance frameworks, improved corporate accountability, and regulatory compliance through timely, accurate, and detailed reporting on the business performance.

- Ensuring security and privacy of customer data as the business processes change, old data becomes redundant, and new data is brought in.
- Managing environmental responsibilities with Lean and efficient business processes, efficient data centers, and sustainable human resource (HR) policies.
- Enhancing electronic presence through social media by exposing the right areas of the organization to customers, potential customers, and business partners.

Managing human behaviors

People are involved in all business functions, including financial management, customer relationship management, and supply chain management. People are also involved in HR (people management). Each business function requires decisions, and data analytics provides the ability to make those decisions faster and cheaper. Creating an environment for data-driven decisions requires each user (staff) to understand the importance of the source, storage, and usage of data. An important source of data is the user themselves. Understanding and managing human behavior in the adoption of data-driven culture is the leadership responsibility. Leaders understand that it is the combination of the environment they create and the behaviors they support, which will result in a successful outcome for BO.

Leaders create situations that make it easier for the right people to take the right decisions. Confidence in the use of analytics in decision-making is a multipronged approach requiring demonstration of successful decisions across the organization. For example, first-time success in approving a loan by a customer-facing staff in a financial institution is a story promoted by the leadership across the institution. This promotion creates the necessary impetus for other decision-makers to start using the AI solutions and provide faster and more accurate service to the customers.

As a part of managing the change in the business, leaders also have to handle ambiguity and confusion. Anticipating ambiguity and confusion is important – for no matter how sophisticated the AI solution is, its usage by people is bound to have elements of ambiguity and confusion.

The subjective aspect of data usage by people includes costs, regulatory, security, privacy, and reliability issues. People are interested in the source of data, how much it costs them, whether the data and analytics are protected by regulatory compliance as well as technical security, and whether the analytics will jeopardize their privacy. People work on trust and reliability of the analytics they are using. Large social media organizations acquire data directly from users and then they further provide it to other vendors in the analytical market through Analytics as a Service (AaaS) to others. AaaS can be a cause for concern for end-user customers who need assurance of security, privacy, and reliability of the services.

Human behavior is also a factor of perception. Therefore, optimization relies heavily on the quality of information and insights generated by AI in a business process. In particular, quality needs attention across business silos that include both internal and external stakeholders. Chapter 8 further focuses on these important aspects of quality.

Sociocultural factors impact the implementation of AI and the transformation of workers to knowledge workers. Staff manage the business risk as services are automated and optimized. But staff also face the risks emanating from AI adoption. For example, as the number of people required to execute a process reduces, the staff can perceive that as a threat to their jobs. Cost savings and business efficiency alone are unlikely to drive the enterprise-wide acceptance of AI-enabled services. Therefore, cost savings associated with an activity are less important than the acceptance and use of a process by people.

Incumbency is highlighted as a significant people-based business risk during BO. Automation may not always result in acceptance by people. This is mainly because automation introduces uncertainty in the minds of staff in terms of their own future. Optimization further exacerbates this challenge of uncertainty in the minds of users.

Regulation is required to ensure the security and privacy of data. Regulations, however, can also contribute to the slower adoption of AI-based decision-making across multiple industries. For example, data breaches in a taxi company can lead to slow adoption of AI in an airline company. Collaborations across multiple industries lead to the challenge of regulations impacting adoption.

Due consideration to these soft (people) factors is required and that is a leadership responsibility.

HUMAN RESOURCE (HR) MANAGEMENT

The Human Resource (HR) function of an organization has to handle two distinct aspects of culture change due to BO. One is the handling of change across the organization. The second is changes to processes and operations of departments with the adoption of AI. HR is responsible for revised processes and services. However, HR processes based on cloud services (e.g., Software as a Service or Infrastructure as a Service) also change with AI: for example, the upgrading of data science capabilities through internal training, the process of recruitment, and the compensation and reward structure – these all change with AI adoption.

HR process changes

The following HR processes change with AI:

- Performance metrics to monitor, reward, and promote people are optimized with data analytics.
- The recruitment process eliminates the intermediary processes and directly uses analytics-based profiling.
- Redefining roles and responsibilities due to changes from BO and managing transition of roles. This will include mentoring and coaching for AI-enabled decision-making. Existing expertise in the organization can be upskilled to use AI in business processes. Once developed, this same expertise is used in mentoring other staff.
- Defining new roles in BO that will provide data analytics and that will learn to use data analytics in decision-making.
- Anticipating changes to behaviors in HR personnel resulting from optimization.
- Enhancing professional skills development through training, mentoring, and recruiting. This would be customizing training based on the competency level of the teams within the program of work. Train and mentor champions to promote AI-based processes. Invite these experts to provide training.
- Training for the entire end-to-end business processes. This training starts with the key goal of the process and then helps the users understand the activities and tasks impacted by AI.

Organizational process changes

The following organizational processes change with AI adoption:

- Modeling processes by investigating their activities, roles, deliverables, and practices. Identifying the diverse pull for each of these process elements by the corresponding methods, then work to reduce the effect of this handling that methods friction and pull in separate directions.
- Involve end-users and sponsors in the entire adoption effort so they are able to anticipate and participate in the change arising in the processes. The iterative and incremental approach to adoption ensures that the users have the right expectation from the changes.
- AI presents new business opportunities and challenges. Business professionals need training and mentoring to handle the challenge.
- AI and networking increase the opportunities to work remotely. This change creates newer roles and challenges of handling' privacy and security of processes
- Data analytics, visualization, risk, and security are all included in the soft issues of culture that impact employees, suppliers, customers, and shareholders
- Communicate across the organization the need to transform with the adoption of AI. Identify the elements from the higher-level (business-level)

processes that can benefit from AI. Embed AI-based within those higher-level processes.

- Align processes with each other by reducing "methods friction."[3] This is called "process alignment."
- Align the business internally to its existing technologies and systems and externally to its business partners and customers.

Virtual and collaborative teams

The creation of virtual teams based on the niche skills of various employees, consultants, and managers can lead to multiple offerings by the business as various players can get together to serve the needs of a particular customer. Such virtual teams enable the business to tap into the skills of consulting professionals outside the business for shorter and specific durations. While the purpose of such reengineering of processes is not staff reduction, they do lead to a much more Lean-Agile team structure.

Training business people

Business people (nontechnical) need training on ongoing basis to help them utilize optimized business processes. Three questions non-tech employees should be able to answer[4]: How does AI work? What is it good at? And what should it never do? This requires training and mentoring for business users to have a good understanding of the basics of AI. Second, the business user is trained to spot the activities and tasks that are optimized through AI. And third, the users are trained on the security and privacy of AI in their own processes.

Based on Knickrehm,[5] the following needs to be planned in changing the culture of the organization:

- Augment human skills in business with AI rather than aiming to replace humans altogether. Total replacement of humans is envisaged in pure automation; but in BO, humans are complimented by AI
- Reinvent operating models and reengineer processes in order to accommodate data-driven decision-making. Business process reengineering discussed in the previous chapter deals with the creation of newer models of business and its processes. These processes are aimed at providing value to customers with efficiency and effectiveness.
- Redefine the jobs for the staff and associates which will accurately reflect their ability to make business decisions using AI-based insights.
- Rethink organizational design to a flatter, collaborative organizational structure rather than a hierarchical one.
- Encourage staff to partner in the new organizational initiative rather than being "told" to do what they are supposed to do.

Ascertaining current skills and competencies in the organization and planning for upskilling can be based on frameworks such as the Skills Framework for Information Age[6] (SFIA). SFIA can be used across multiple business processes with the organization.

Educating the customer

Customer value is enhanced by direct and accurate access to the organization and its services. This access transforms the business into an "available" business that enables the customer to obtain personalized services from the business. For example, an airline passenger with access to the flight schedule, times, and check-in facility is able to manage himself or herself much better than with the corresponding manual flying procedures. Similarly, accessibility to multiple sources of information such as weather, road conditions, and news reports can all add up to a pleasurable and safe travel experience.

In addition to training and upskilling the staff and associates who are providing the service on behalf of the organization, it is also important to pay attention to the changes that the customer will experience as a result of AI-based decisions. For example, the most change that customers experience is a chatbot using IVR (interactive voice response) to answer their questions. Customers can also be educated to use "Self-Serve Analytics" to figure out the basic, automated answers to queries on loans, credit card facilities, airline prices, and hospital bookings. Customers weigh in the advantages of using Self-Serve Analytics, providing they are not overwhelming. Customers also interface with the business through sites, such as Twitter and Facebook, enabling the organizational systems to interact with external systems.

- Employees – Organization moves to a lean structure. High value-adding permanent staff to support core functions and drive business transformation and services. This is supplemented by managed vendors and a resource pool of skilled specialists brought in as required through an optimized recruitment process.
- Rewards for Innovation – A suitable reward and recognition values framework is required across the enterprise.
- Customer Focus – Business areas, including technology, are often internally focused with "people operating around them." A move to a more customer-centric focus is required.
- Management Capacity – The ability of an area of business to undergo change is often limited by the capacities and capabilities of management. Building the soft skills of management is part of the successful transition.
- Vendors play a significant role in managing public and private cloud-based vendor management as part of managing business change.

ADOPTING AI FOR AN AGILE CULTURE

Agile is a culture in its own right. Businesses strive for quality to maximize customer value. An agile culture enables businesses to respond effectively to internal and external changes. Agility also initiates changes of its own. AI contributes to agility by providing timely updates on the status of a business to facilitate decision-making.

A flexible business model and associated Agile corporate culture are capable of handling sudden external changes. Accompanying the need for structural flexibility of business is the need for the underpinning systems (e.g., HR, customer relationship management, CRM) to facilitate such nimbleness. BO integrates these technologies and tools with processes and people, thus paving the path for a flexible business structure.

BO changes an organization's internal operating structure, alters its relationship with external parties, and affects the business ecosystem. BO crosses the boundaries of systems and processes resulting in agility.

Embedding AI in business processes changes the way the entire business operates. Decentralized command and control with distributed operations enables a maximum number of staff to make decisions. Changes are brought about in development and maintenance, budgets and costs, support and services, architecture and infrastructure, and legal and environmental aspects of business. Sociocultural, psychological, and motivational issues are part of leadership challenges in BO. Systems Thinking, as described by Senge in Fifth Discipline,[7] helps in managing these changes. Unhelkar (2010a)[8] has further argued for the need to handle people issues by getting together project sponsors, business stakeholders, customers, and users, along with architects, designers, developers, and testers, to implement solutions.

The agility of business depends on the type and size of business. For example, a small travel company can adopt agility in its operational processes using basic data analytics, whereas a large hospital transforms all its business functions to Agile. Thomsett (2010)[9] advises beginning any business transformation with a series of "fierce" conversations around culture, values, and behaviors. Such conversations, he argues, will ensure that the people are totally involved and are fully aware of the potential cultural impacts. BO benefits such open and robust conversations. Eventually, BO makes an organization agile, customer-driven, and an enabler of personalized services. Following are the advantages of AI to customers on the sociocultural level:

- Enhances customer experience through personalized and location-specific services due to fine granular analytics.
- Provides customers with a range of additional products and services due to collaborating businesses.
- Provides customers with the ability to demand rapid changes to their existing orders with reduced risks.

- Enables a customer business to quickly initiate changes to the existing order.
- Facilitates ease of compliance for the customer, and this is achieved by providing the customer with timely data on regulations, securely and privately.
- Extends customer global reach through wide offerings of products and services with the use of communications technologies and Web services.
- Improves the quality and efficiency of products and services.

The sociocultural environment of the organization becomes more agile and flexible as AI is adopted by the organization. For example, with AI, the demographics of customers on social media are better understood. These analytics provide valuable information on customer sentiments and their trends. The organization can respond in anticipation, making it agile.

Adopting AI also changes the way goods and services are sold online. Agility with collaboration expands the reach of the organization beyond its geographical boundaries. The expanded reach provides agility to the business to serve wider customer needs.

Agility is also the ability of a business to creatively generate new products and services, come up with innovative ways of handling the competition, and prioritizing its risks. An Agile business creates many opportunities due to its creative and innovative culture. Enabling innovative approach to business often calls for changes in business practices and business operations. The need to foster an innovative culture is also high in BO, which enables people to experiment with processes and technologies to improve and optimize them.

The Lean Six Sigma approach in the business methods space, or even governance standards, such as the Information Technology Infrastructure Library (ITIL),[10] can be customized to carry out BO. These process frameworks can enable BO by capitalizing on the people, processes, and technologies of the organization.

Agile culture is also a cybersecurity-aware culture. Agility is flexibility without sacrificing the internal and external security, and also its physical and electronic aspects.

Successful transformation underpins the principles and practices of a legal framework that can be used to understand the contractual obligations of the organization, particularly related to electronic transactions arising from collaborative commerce. This is particularly important as AI Internet technologies are embedded within processes.

CONSOLIDATION WORKSHOP

1. Discuss the importance of leadership and culture change in BO. Argue why having good leadership is vital in the adoption of AI for BO.

2. What are some of the salient characteristics of astute leadership in the context of changes to the mindset of users?
3. How can leadership help in automation versus optimization?
4. Why is it important to manage change in BO? What will happen if AI is treated as a technical project and culture change is not managed?
5. What are some training (and ongoing training) mechanisms to bring about a change in mindset?
6. What are the advantages of understanding organizational cultures and business agility in an AI-based transformation?
7. Who are the key stakeholders in a business and what role do they play in AI adoption and change?
8. How do HR departments impact agility in AI-enabled businesses? Discuss this in the context of the changes to the HR processes and the changes to the organizational processes supported by HR.
9. What are the nuances of handling the changes due to collaborations?
10. Is there a need to educate the customer when the business is being optimized? What can be done to educate the customer easily?

NOTES

1. Sherringham, K., & Unhelkar, B., (2020), *Crafting and Shaping Knowledge Worker Services in the Information Economy, Springer Nature* (Palgrave Macmillan), Singapore, 2019. ISBN 978-981-15-1223-0; Ch 7.
2. See note 2.
3. Unhelkar, B., (2012, 20 Aug), Avoiding method friction: A CAMS-based perspective – Cutter Executive Report, Boston, MA. Vol 13, No 6, *Agile Product and Project Management Practice.*
4. Martinho-Truswell, E., (2018) "Three questions non-tech employees should be able to answer" *Harvard Business Review*, Boston, MA, 65–78.
5. Knickrehm, M., (2019), "How will AI Change work?" Artificial Intelligence, *Harvard Business Review*, Boston, MA, 104–106.
6. von Konsky, B. R., Miller, C., & Jones, A., (2016), The skills framework for the information age: engaging stakeholders in curriculum design. *Journal of Information Systems Education*, 27(1), 37.
7. Senge, P. M., (2006, Jan 1), *Fifth Discipline' The Art & Practice of the Learning Organization*, Broadway Business; Revised edition.
8. Unhelkar, B., (2009) Agile as a psychosocial paradigm: CAMS fundamentals beyond the manifesto – Cutter Executive Update, 1 of 5, Vol. 14, No. 10, Agile Product & Project Management Practice, Boston, MA.
9. Thomsett, R., (2010), Agile sponsorship: the next element in the Agile evolution, Cutter Executive Report, 1 April 2009. Boston, MA.
10. Information Technology Infrastructure Library (ITIL) - https://www.axelos.com/certifications/itil-certifications/itil-strategic-leader-itil-4 accessed 2 Nov 2020.

Chapter 8

Quality and risks

Assurance and control in BO

INTRODUCTION

This chapter discusses the quality and privacy of data in optimized digital business processes. The characteristics of Big Data discussed in Chapter 2 (3+1 Vs) present unique challenges in terms of quality.

Big Data characteristics include lack of format, schemaless-ness, high volume, and high velocity – each of these characteristics leads to a challenge in handling the quality. The user communities on social media produce alternative data. These are the customers solving their own problems and helping each other via communities. This data is not owned by the organization analyzing it. There is no opportunity to verify the authenticity of such alternative data. The question of privacy is continually challenging the use of such data. Additionally, artificial intelligence (AI) solutions are exposed to data without initially establishing the context. Context around a data point is crucial for quality, as it provides the basis for analytics and subsequent testing. Therefore, simply testing an AI solution thoroughly using a range of data is not enough. *Think data* (Chapter 2) and all four of its aspects come into play in ensuring holistic quality. Governance, risk, and compliance (GRC) provide necessary controls to enhance the quality of decisions and reduce the risks associated with data usage.

Quality initiatives establish trust and reliability of AI-based solutions in the minds of the users. Robust quality approaches to AI go beyond the technological quality of the solution and provide business benefits. Coupled with robust governance frameworks, visibility of quality and services helps in establishing business value.

Quality impacts business decisions. Relying on data analytics in decision-making depends on trust. Insights from analytics are of value only when the perception of quality and service assurance are met. The size, speed, and variety of data and the corresponding analytics become irrelevant if the users start doubting the analytics outputs. Quality is an overall function to establish that trust by preventing and detecting errors well in advance and resolving the problems that occur in usage.

Quality initiatives include validating and verifying the quality of data, analytics (intelligence), model, processes, and usability. These quality initiatives are well supported by testing the source of data, applying feedback and results iteratively, and providing high visibility of the changes. Managing the intricacy associated with data acquisition, storage, cleansing, usage, and retirement is included in this all-encompassing aspect of quality.

Database systems, information systems, and knowledge-based systems could function independently before BO. The systems, hidden behind the firewalls and in silos, could not be easily hacked. Thus, by their very nature, pre-AI application in BO, enterprise systems were relatively secure. BO requires systems to work together. Customer call centers, enterprise web pages, online shopping stores, banking, and e-commerce cannot work in silos. Analytics and services are based on an interconnected world. This interconnectedness leads to quality and security issues. These issues are of overall quality environment and cannot be solved through testing alone.

The quality function includes quality of data, analytics, models, algorithms, code, processes, and people. Quality is also a recursive function as it is also responsible for itself. The quality management function is itself subject to the quality criteria of the quality environment. Tools and techniques related to quality and testing support the quality process. Transitions, training, project selections, prioritizations, and documentation are quality issues associated with BO. The quality environment is an iterative and incrementally operating environment that contributes directly to the value-generating effort of BO.

Two practical aspects of quality are testing and assurance. These quality aspects rest on detection and prevention. Detection occurs through testing and prevention is through assurance. Quality control deals with the detection of errors. This is also called testing. Testing focuses on identifying errors as compared with a reference point or an ideal situation. Testing requires a valid data, and a model against which the new artifact being tested is judged for its quality.

Quality assurance primarily deals with the prevention of errors so as to provide an excellent user experience. This assurance is the discipline of quality processes and models to ensure the prevention of semantic and aesthetic issues. The aim of assurance is an end product that is free from defects.

Quality assurance in BO includes activities such as setting data filters at the source of the data, proactively managing the meta-data in addition to the transactional data, and use of agile iterations and increments in developing AI solutions. Quality control or testing examples include the data and verification of algorithms, testing the nonfunctional or performance aspect of the solution, and fixing the "bugs" and retesting.

Direct and indirect impact of bad quality

There are numerous direct and indirect impacts of bad data quality on business decision-making. These impacts of quality specific to business optimization are as follows:

- Direct, short-term tactical operational decisions are affected by poor quality of data. Defective products get installed and used by customer. Other examples include accounting (payment and invoicing) errors. These errors are rectified by undertaking a data cleansing and testing exercise.
- Direct, long-term strategic decisions are impacted by poor data and analytics: for example, loss of partnership or acquiring wrong partners. These decisions cannot be easily corrected by simply correcting the data; all underlying business processes need to be revisited, and so also the algorithms analyzing the data – in order to improve the quality of these decisions. The more strategic is the decision, the greater the need for agility.
- Indirect, short-term tactical operational decisions have poor quality: for example, poor inventory management, slow production, or poor customer service that are hidden in the processes. Handling these indirect poor decisions requires improvement in data and analytical quality, coupled with coaching and training for the staff and customers using the insights to make those decisions.
- Indirect, long-term strategic decisions are impacted due to poor historical data and no control over alternative data: for example, the development of a completely wrong product or inefficient and undirected marketing, compliance issues, and loss of goodwill of the organization in the market. The quality situations can be improved by a holistic, strategic approach to quality assurance that includes improving data, information, algorithm, knowledge, and decision – coupled with enhancing people skills, attitudes, and influences.

Risks and governance policies

Risk management closely accompanies quality initiatives. Governance, risk, and compliance (GRC) provide the framework to control risks and ensure compliance. Quality is further augmented by GRC, which enables the creation of an overall quality environment that is strongly focused on prevention rather than detection of errors. A GRC initiative is thus also useful in creating services quality.

Good governance is a balancing act. While maintaining required business outcomes, it also needs to ensure that the business is not so overloaded that it ceases to exist. Pragmatic governance balances risks with opportunities, cost, and the need to deliver services and operations. An important responsibility of risk management[1] is to keep business viable. Governance should not negatively impact quality, cost, and the operational efficiency of the organization.[2]

GRC is a part of the quality environment. The GRC effectiveness and efficiency are enhanced by understanding the business environment, type and size of business, existing capabilities and limitations of the business, and the available tools and technologies for the governance function. Furthermore, due to the prominence of compliance acts, such as the Sarbanes–Oxley Act (SOX)[3] and Health Insurance Portability and Accountability Act (HIPAA),[4] AI applications are required to provide explanations for the decisions arrived at. Regulations also make it mandatory to provide auditability and accountability of AI-based solutions.

General data protection regulation (GDPR)

General Data Protection Regulation (GDPR)[5] dictates rules for the collection and processing of personal data. The regulation also provides the user with the right to port their data from one organization to another – such as a patient moving from one hospital to another.

While GDPR jurisdiction is the EU (European Union), it has a much wider applicability in the digital world that transcends geographical boundaries. GDPR enforces organizations to justify their collection of data and explain the reasons and methods of its processing. Electronic data cannot be collected without unambiguous and explicit consent from the user. Furthermore, the business must provide the users with a copy of their own records, correct the details therein when asked by the user, and erase the data if the user wishes so.

Quality and ethics

The ethical, social, and legal consequences of decisions impact the quality of BO. These issues cannot be handled by AI and its automated ML systems on their own. AI systems do not understand the context in which decisions are made. Therefore, it is crucial for an enterprise to validate the quality of predictions and decisions based on AI. Chapter 10 describes a model of combining human natural intelligence (NI), experience, and intuition in every stage of the automated ML decision-making pipeline. Human subjectivity which is an indispensable parameter for producing and validating quality decisions and customer value plays an important complementary role in supporting and enhancing BO with AI solutions.

Big Data-specific challenges to quality and testing

The following are Big Data–specific challenges to quality and testing:

- Analytics use a wide variety of structured and unstructured data that require different testing techniques to ensure their quality. Structured, transactional data can be subjected to traditional testing techniques based on equivalence partitioning and boundary values. Unstructured, schemaless data cannot be tested with the same techniques. The automated data preprocessing and the dynamic online data monitoring (AI tools and frameworks presented in Chapters 4 and 5, respectively) are ideal for quality assurance and control of this kind of unstructured data.
- The data quality practices include data profiling that is based on the potential users. Data profiling maps the data to the desired outcomes. Therefore, in a way, the data profile starts to provide the context of data usage.
- Changes to the context in which data is used are based on the changes in outcomes desired. This context change results in different perceptions of the same dataset and, therefore, requires a different quality strategy (including testing) for the same dataset. For example, a dataset on interest rates has to be tested for algorithms predicting changes to the interest rate versus predicting risks to the loan amount.
- Changing levels of granularity. Analytics will have to be tested against a range of granularity, and that can be challenging. Fine granular analytics needs to be tested with high-velocity data on a very narrow requirement.
- Externally created and externally sourced data cannot be easily filtered for quality. Furthermore, should a filter spot an error, there are very limited means available to correct such data. And on correction, that data may end up with a different format to the original data – creating challenges of inconsistent data formats. Externally, data that is not relevant to the business context or one whose source is dubious must be kept outside the firewall.
- The need to continuously and rapidly align the new, incoming Big Data with existing transactional data. New data potentially contain anomalies and security risks that are identified only when an attempt is made to align it to the existing data.
- The need to test the concurrency of data processing that requires the balancing of loads to ensure operational performance.
- Operational parameters can wildly vary. They may not always provide the opportunity for a satisfactory user experience (in terms of time and visuals).
- Cybersecurity of data that is spread within and outside the organization is a major challenge. Strategize for use of security analytics tools.

QUALITY OF "DATA TO DECISIONS"

The pyramid shown in Figure 8.1 summarizes the various items, processes, and their relationships in the business world. Data management, data quality, data consistency, data access, data continuity, data completeness, and data permissions form the basis for data quality. Each layer in the evolution of data to decisions requires attention to quality.

The pyramid layers based on earlier discussion in Chapter 2 are stacked from bottom to top with data, information, services, knowledge, and decisions. Traditionally, business organization is set up hierarchically – with teams engaged with the duties in the lower rungs of the pyramid, while the managers occupied the higher rungs, chiefly as knowledge heads involved in decision-making. Data contains raw figures obtained from sensor measurements. Information is the interpretation and meaning imposed on the data by humans. Services are orchestration of various information systems at a higher level. Knowledge is the association of various types of information and services making action possible. Decisions deal with the selection of a particular course of action, based on knowledge.

AI requires businesses to change from the traditional pyramid hierarchy. AI and ML have been extensively applied to business, and closer inspection reveals that AI and ML algorithms limited to the bottom and top layers of the pyramid is a lost opportunity. Large quantities of data (bottom layer) are collected, preprocessed, and directly fed into ML algorithms, the result of which are predictions and/or decisions (top layer).

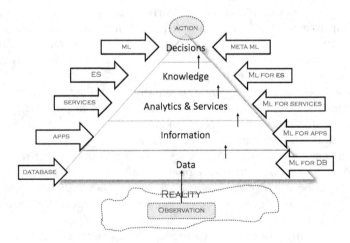

Figure 8.1 AI impacts and is impacted by the evolution of data from observations to decisions.

Quality of data

Data is the first layer of quality that ascertains its veracity. Quality data is essential for quality AI and ML. Techniques such as equivalence partitioning and boundary values can test the veracity of data. These techniques make use of sampling, checking, and correcting the data based on the parameters provided by the business. The data quality initiatives surrounding Big Data need to handle factors in addition to testing just the "data." For example, large volumes of unstructured data simply cannot be tested on their own. An understanding of the context enhances the value of data analytics/semantics before the data can be tested.

The peculiar nuances of Big Data bring additional challenges to quality. For example, consider how data is viewed as an "aggregate" in the unstructured data. This viewing of an aggregate implies a move away from the structured rows and columns of a relational database. Testing and ensuring the quality of such unstructured data cannot be undertaken by the traditional sampling from a set of data and testing it. Sampling of data based on equivalence partitioning and boundary values presumes a semblance of structure within the underlying data. Such structures are often not available in Big Data. Hence, testing of Big Data sets may have to occur over an entire data set rather than a sample.

Another important consideration in testing and quality assurance of Big Data is that the data on its own may have very limited parameters that can be tested. For example, basic filtering of input data can ensure that numbers and texts are in their respective fields within a form on a Web page. Beyond that, basic filtering may not be able to ascertain and test the semantics behind the number or text. Additionally, with Big Data, the data values may not have a format, and therefore even the format-level filtering may not apply.

Merged data used for the calculation of derived values includes account management, permissions management, database indexing, and log file management. Testing this data requires an integrated test strategy with the required functional testing, user acceptance testing, penetration testing, accessibility testing, and load testing.

In addition to the sheer size of data, there are other factors impacting the quality of data. These factors include infiltration of missing, misplaced, and distorted datapoints. ML tools can also be used to address the data quality problem along the following lines:

- Missing data is a common problem in many datasets. A couple of attributes in a couple of records or some entire records could be missing in a dataset for several reasons, including human error. ML algorithms trained on relevant datasets can easily detect the missing datapoints and fill in appropriate values obtained through the learning process.

- Data augmentation techniques include ML to solve the problem of insufficient data in numerical, text, and image data formats (Chapter 4). In a similar way, ML algorithms look for distorted or inappropriate datapoints.
- Numerical data in which values are misplaced or distorted or out of range are very difficult to detect when dealing with large datasets. For example, the data entry of someone's year of birth as 1794 (instead of 1974) goes unnoticed when data scientists must check tens of thousands of data records. ML algorithms trained on the validity of datasets can pinpoint incorrect values of data attributes and, in some cases, even suggest solutions.
- Text data product reviews by customers, customer sentiment analysis, blogs, and so on are the modern sources of data in the text format. Misspellings, wrong usage of words, and grammatical and syntactical errors can distort the meaning of sentences or make them meaningless. Conveying correct syntax and semantic meaning is important because text-based information is the source of knowledge. For example, in the course of typing the contents of this book, several typographical errors like *machine leaning* (instead of *learning*) go unnoticed. Such subtle errors in text data can be traced and fixed by ML algorithms trained on large language corpora.
- Image data. Human eyes soon get exhausted looking at the details in colored images. It is humanly impossible to check millions of images that are routinely crunched by image processing programs in computer vision-related tasks. In addition, some of the latest security attacks come from Generative Adversarial Networks (GAN),[6] which fool even the state-of-the-art image recognition algorithms. For example, introducing an infinitesimal perturbation at the right intersection of pixels in the image of a panda can mislead even the highest performing algorithm to recognize the panda image as nematode or gibbon with exceedingly great confidence, even though nothing has changed on the surface when viewed by human eyes.[7] ML algorithms trained with GANs are capable of detecting such errors.

Quality of information

Level 2 in Figure 8.1 represents the processing of data to create information. Information is a systematic identification of patterns and trends within those data. Data, on its own, is not meaningful, whereas information based on the data provides meaning or semantics. Ensuring the authenticity of this meaning is the responsibility of quality of information. The e-world comprises billions of pages that are increasing exponentially. There is no centralized agency monitoring the quality of information available on the internet. Individual information systems on the web are constantly trying to adjust to the flow of information on the web so as not to be outmoded. It is difficult to test the quality of the information fed into these systems from the web. ML algorithms placed at the front-end of web-based information can learn to test and validate the quality of the incoming information. News, podcasts, articles, blogs, auctions, advertisements, shopping sites, and mobile apps are subject to testing via ML algorithms.

A caveat in the BO world is that anomaly in information is not always a lack of quality. Anomalies can signal security breaches which require detection and filtering. Anomalies can also indicate emergence of a new idea. For example, consider how spam filters in late 2019 filtered out mails based on certain keywords like "virus, infection, epidemic, disease, and so on." Filters embedded with ML functions would have possibly noticed the emergence of a new entity (like COVID-19) by analyzing the anomalies in the email information.

Quality of analytics and services (collaborations)

Analytics quality is the quality of its algorithms. Analytics are complex, and their algorithms need to be verified from both statistical techniques and a programming viewpoint, including exceptions and error handling. The quality of analytics includes verification and validation of their syntax, logic, standards (e.g., naming of attributes and operations), and the purpose they serve (semantics). Analytics establish correlations between data and information. The more the data, the better is the output. Also, data analytics is not limited to analytics on a singular type of data. Data is sourced with the help of services from multiple and widely varied databases (typically on the Cloud) and a relationship established between them in order to perform data analytics. Analytics itself is offered as a Service[8] on the Cloud.

Since the algorithmic code deals with and manipulates the data, the quality of that code also influences the quality of the data. A white box method for quality of analytics approach to quality of analytics consists in meticulously checking the code for syntax and semantic errors. White box methods can be tiresome and not foolproof. ML offers a black box approach to quality testing of analytics and corresponding services. Continuous testing based on Agile iterations enable black box testing. An ML algorithm provided with known output for a given analytics and services can learn (in a supervised way) to detect errors and faults in similar analytics and services on the web.

These tests (quality control) are also meant to detect performance and reliability issues. Quality assurance of the algorithms happens through models and architectures and following a development process (i.e., Agile in the solution space).

Self-serve analytics (SSA) require the sharing of data across the enterprise, including partners, customers, and providers. The quality of SSA can depend on a number of factors, such as the ease of configurability of the analytics, the expertise and experience of the user, and the urgency and criticality of the analytics. Most of these factors are outside the control of the organization, and testing them requires considerable assumptions. These assumptions include the reliability of the source of analytics, their compliance with the legal requirements of their jurisdiction, and the use of SSA by other business partners.

Data usability is the ease of use of data within analytics. The accessibility, profiling, cleansing, and staging of data play a part in data usability and, in turn, its quality. These factors affect the quality of the solution being developed.

Quality of knowledge and insights

Knowledge is rationalization and correlation of information through reflection, learning, and logical reasoning. Analytics form the basis of such correlation. The body of explicit knowledge available in an enterprise along with the implicit knowledge in the form of intuition and experience of business experts is optimized in BO. End-users and domain experts seek this knowledge.

Use of knowledge and insights is based on analytics from expert systems. These expert systems in the past were insular, with the body of knowledge occasionally updated to keep abreast with the latest domain knowledge. ML-based systems analyze and filter online information and engage in knowledge discovery. Newly discovered knowledge is automatically added to the knowledge base. Increasing the reliability requires testing the discovery capabilities through test data and on a real-time basis. Tested results are inputted via feedback loop back into the system and further tested to enhance its capabilities.

Quality of decisions

While data, information, analytics, and (to a large extent) knowledge are considered objective, decisions are a human subjective trait. Quality of decision-making depends on tacit human factors such as personal experience, value system, time and location of decision-making, sociocultural environment, and ability to make estimates and take risks.

Quality of decision is a result of the quality of the previous layers shown in Figure 8.1 and combined with NI. AI-based systems can suffer data bias, algorithm bias, and decision bias. Large amounts of skewed input data introduce data bias. Algorithms can be designed by developers with preconceived notions that can be biased. Eventually, decisions based on data and algorithm biases can themselves be biased. The feedback of consequences from these decisions creates further biases. Therefore, the quality of decisions can be improved by iterative feedback, by evaluation of consequences by multiple people and by keeping mind the context, time and place of decisions.

Finally, intelligence is actionable knowledge based on insightful use of Big Data. The decision-maker's ability to distinguish decisions based on their important, relevance, context, and organizational principles is a crucial quality factor for intelligent decision-making.

Testing is performed at best with software tools. The ML technology discussed here is applied to assure quality of the middle layers of the data-decision pyramid. ML can be used as a tool to test and refine the quality. ML algorithms need the application of quality techniques to enhance their quality.

QUALITY ENVIRONMENT IN AI AND ML

Assuring ML quality

Analytics can be used to ascertain the quality of data. This data and the corresponding analytical algorithms are subjected to quality assurance activities using the capabilities of AI. For example, an analytical algorithm can be applied to the incoming weather data to identify potentially unrealistic spikes (e.g., a temperature variation of 500°F at the same location within a few minutes). Such identification of spikes can lead to an investigation of the algorithm itself to ensure that it has passed the testing and is secured.

Quality assurance and testing of ML algorithms/systems are carried on at three levels:

1. Testing performance and logic of ML algorithms
2. Testing for bias in the ML predictions
3. Reducing the inexplicability of ML algorithms

The first is discussed here, and the second and third are discussed in Chapter 10.

Testing of ML algorithms starts by separating the data into testing and training. Testing the algorithms also requires a strategic approach including black- and white-box testing. Unstructured data, in particular, is challenging to test. Creating sample data and testing the execution of logic are recommended. When the questions themselves are not known and the correlations are produced by the machines, quality assurance becomes a risk management exercise.

The opportunity for feedback loop in testing Big Data and its analytics is much less. This is so because the topography of themes provided by the Big Data is not known at the onset. Testing requires validation of data and algorithms. Testing of BD systems requires a certain amount of guesswork.

The quality of data and the validation of algorithms are never entirely complete. Quality assurance and testing of ML systems require validating the test data, validating the training data, and then validating the performance of the algorithms against open data.

GRC positively influences the transformation and delivery of BO. Leadership defines, leads, and manages BO and its overall quality. Leaders not only set the objective and strategy, but they also set parameters for quality. Leaders balance the cost of making bad quality decisions and also not making decisions. BO comes with a risk and the risk is higher when quality of data and algorithm is not verified and people are involved.

The quality and value from Big Data are also based on the perception of the users of the analytics and business processes. Therefore, ensuring the quality of Big Data goes beyond technologies and also includes sociology, presentations, and user experience. These are some important aspects of quality that are specific to Big Data and that need to be handled on a continuous basis for the analytics and business processes to provide the necessary confidence and value in business decision-making.

Assuring quality of business processes

The quality activities here include modeling of business processes and, thereafter, the verification and validation of those business processes using techniques similar to those used in the quality assurance of models and architectures. Process quality depends on the way in which the processes are executed by the users and their end goals.

The quality of business processes is verified and validated by creating visual models and then applying the quality techniques of walk-throughs, inspections, and audits. Business processes use applications and analytics to help the end users achieve their goals. Therefore, the quality of business processes depends on the way the users perceive their achievements. Understanding, documenting, and presenting the visual models of the business processes to the end users and incorporating their feedback in an iterative manner are Agile ways to enhance the quality of business processes. Process modeling standards (such as Unified Modeling Language [UML][9,10,11] and Business Process Model and Notation [BPMN][12]) and corresponding modeling tools further help in improving the quality of business processes. In addition to the processes that form part of the business, there are the processes that deal with producing the solutions. Project management, business analysis, and solutions development life cycles are examples of these processes.

The quality of these adoption and solutions development processes is also important and needs to be subject to the same quality techniques as those used for quality in business processes. A set of well-thought-out activities and tasks combined with the Agile techniques produce accurate and higher-quality analytics which, in turn, enhances the quality of the business process.

Developing the quality environment

The following are the strategic considerations in developing a quality environment in a Big Data initiative:

- Identifying key business outcomes: Defined in the corporate strategy in order to create a common understanding of the Big Data adoption exercise and the role of quality in helping achieve those outcomes. These business outcomes have a bearing on each of the data quality characteristics. The more detailed the outcomes and the higher the risks associated with those outcomes, the greater is the demand on data quality.
- Modeling the range or coverage of Big Data: Its sources, types, storage mechanisms, and costs. Clarity in understanding the range of data enables the formulation of quality strategies based on the depth and breadth of the incoming data, associated risks, and costs associated with analyzing that data.
- Extent of tool usage: Most Big Data quality initiatives need tools and technologies that complement the technologies of Big Data. For example, quality assurance and control activities on a NoSQL database will need tools that can verify the extraction of data, match the extracted data against reference data, and provide feedback to the quality personnel. This verification exercise can be challenging because of the unstructured nature of the data; therefore, tools that can handle the testing of unstructured data and its performance are required. Tools are also a must because of the high velocity of incoming data and the varying levels of granularity in analyzing that data. Quality tools need to be able to operate within the Big Data environment.
- Ensuring a balance between the rigors of quality and corresponding value: The business decision-makers need to collaborate with the quality personnel to ensure that quality efforts are balanced with the business outcomes. Standardization of data and its cleansing in order to enable processing can either go overboard or be carried out over data that may not be used at all. Therefore, it is important to keep the ultimate usage of the data and the desired business outcomes in mind before undertaking quality activities on the data

Assurance activities

Quality activities are carried out over the key phases that are transitioned by data entry, storage, cleansing, and retiring. In each of these phases, the volume, velocity, and variety of Big Data (including myriad data sources) add to the challenge of data quality. These challenges include complications of data governance and risks associated with the management of data. Big Data in particular needs continuous filtering and standardization. Following are the data assurance activities for Big Data:

- The entry point for Big Data has "presumptions" in their use. These presumptions are needed to create and apply filters to that data. These presumptions can be based on prefabricated (i.e., halfway completed) analytics. The fuzzy and uncertain nature of the use of the data presents input filtering challenges.
- Sources of data (social media, crowd, other systems) each requiring a specific filtering before entry, and each of these data sources is not always under the organization's control.
- Lack of context requires presumptions about the use of data before filtering.
- Velocity of data is a big challenge requiring the use of tools for filtering.
- Complexity of analytics applied to the data presents quality challenges.
- Data is identified, secure, and stored – presumably on the Cloud requiring data assurance effort to shift to the Cloud.
- Need to maintain data entities as separately and identifiable as possible due to the 3V of data but the challenge of data quality is further exacerbated when data is mixed types. A NoSQL database requires dynamic modeling and design before data can be stored in it.
- Strategies for backing up and mirroring of databases should be in place. This allows for efficient restoration that is important for data assurance.
- Data is continuously checked for redundancies and abnormalities. Tests are used to remove spikes and prepare the data for staging area where analytics can be performed.
- Ongoing monitoring of data as new data is integrated/ interfaced with existing data for analytics.
- Data retired after use (and when it has lost its currency and relevance) has to be formally archived with the help of tools even in a Hadoop environment.

Developing the testing environment

Testing in the Big Data space requires due consideration to the testing of data, scripts, algorithms, and tools. Following is a list of these testing considerations:

- Testing of algorithms – Development of test harnesses to test algorithms that cannot be executed and tested on their own.
- Test scripts – Writing of test scripts based on use cases in order to test the data algorithms and repeat those tests on a continuous basis in an agile environment.

- Repository of test cases – Which can be used, reused, and augmented as the continuous testing progresses.
- Testing tools – Adopt a standard test tool for test plans, test cases, and results tracking. These tools (e.g. Silk, Chapter 9) also provide security analytics.
- Test planning – A standard process for the creation of test plans and test cases, as well as an outline of the test schedule and resource needs. These plans are based on industry experience, test frameworks, and CAMS agile.
- Test result tracking – Tools and processes for tracking test results and tracking back to requirements and release versions. Analytics on test results indicate areas for rework and regression testing.
- Test approvals – Processes and tools for tracking test approvals and tracking them back to releases and authorizations for releases. These approvals can happen on a Kanban board.
- Testing processes – The required processes, knowledge base, training, and support associated with testing Big Data Testing Types. This is a visible and highly interactive agile process.

Ongoing monitoring of data that is being used for analysis. The quality of data here depends on factors such as changes to the existing data, addition of new data (typical of high-velocity Big Data), and loss of currency of data while it was being used for analytics.

New data is integrated or interfaced with existing data for analytics – and this integration needs to be modeled, tested, and then executed. The integration of Big Data occurs between various data sets (structured and unstructured) owned by the business, data sets provided by external entities (third party), and those being made available through open data initiatives.

ADDITIONAL QUALITY CONSIDERATIONS

All quality efforts are directed at improving the eventual quality of business decisions. Therefore, the basic data quality eventually impacts the business processes. There are a number of quality techniques that are applied at the data and process levels.

- Cleansing and standardizing data – This is a technique to identify and remove the spikes and troughs within a given set of data. In the Big Data domain, this technique becomes important from the point of view of standardizing the data in preparation for its use in analytics. Data editing tools are used in this exercise of cleansing data.
- Applying syntax, semantics, and aesthetic checks to data, algorithms, and code quality. While these techniques are immensely helpful in

ensuring the quality of models and processes, they can also help in reviewing and improving the code and data quality.

- Identifying the source of data and tracing data to that source in order to enable filtering and cleaning – as much as possible. Identification of the source of data may not always be possible beforehand (especially if those sources are identified automatically through a Web service), and in some cases, precise identification of the source may not be permitted (as in the case of identifying the crowdsource).

- Using standard architectural reference models and data patterns in order to provide the basis for the matching of data and thereby spotting mismatches.

- Controlling the business process quality through timely checks and balances at specific activities and steps within the business processes using tools and standards.

- Continuous testing effort as Big Data streaming results in changes to the incoming data points and their context.

- Using Agile techniques, such as showcasing and daily stand-ups, to enable a high level of visibility and feedback in developing high-quality analytics. Some analytics can also be used in identifying errors within a database.

- Using processes and tools for implementing governance policies for data. These can be the automated implementation of electronic policies through algorithms that are specifically created and executed for that purpose. The quality of data stored within the organization boundary can be subjected to greater controls than those acquired from outside. In both cases, though, good data may not always result in good decisions, and vice versa. All the checks and balances cannot guarantee total accuracy. Besides, Big Data input is from sources other than humans. Therefore, cross-checking and filtering out human input is not a guarantee for erroneous data entry (e.g., wrong data created). The greater the number of analyses and manipulation of data, the greater are the chances of loss of quality. This loss of quality of data is primarily felt in the quality of decision-making resulting from that data.

Nonfunctional testing

Slow, unoptimized, and bug-ridden applications suffer a lack of accuracy and performance. Databases, information systems, analytics and services, knowledge-based systems, and decision-aiding systems are composed of AI analytics. These applications need intense quality control of their functionality as well as performance.

Such testing includes performance, volume, and scalability testing. This type of testing is called nonfunctional. The interest here is in the run time performance of the solution (as against its step-by-step function accuracy). This type of testing requires an executable system with fully loaded operational data (or its equivalent synthetic data). An Agile approach to developing solutions helps here because in Agile, testing is a continuous process.

Quality of metadata

Ensuring specific focus on metadata: This focus is on the quality of the context surrounding a data point. Each data point has many contextual parameters that provide additional information about that data point. This additional information, also called metadata, provides a filter for capturing and sharing data elements. For example, metadata around a temperature data point can be the location from where that temperature is being captured. A dramatic change to the next weather data point, in the next minute from the same location, is indicative of bad data quality. The metadata around the collection of data points produces a common reference model that helps filter incoming data. In addition to the quality of the incoming data, there is also a need to ensure the quality of and test the reference model itself. This quality assurance and testing of the model requires cross-functional collaboration, iterative development of prototypes, and incorporation of the feedback back into the metamodel.

In addition to the quality of the data and processes, the quality of metadata is another important element that impacts the quality of business decision-making. Metadata starts to provide an initial context to the incoming data. For example, a tag is basic metadata that is assigned to incoming unstructured data. This tagging provides an identity to the data. This identity and the parameters (metadata) surrounding the data provide a hook for testing – as they improve the chances of filtering out bad data. For example, consider a weather data point showing 800°F. The metadata around this data point provides the location (latitude and longitude) and the time (say, 1:00 p.m.) at which this temperature is recorded. If the next data point shows, say, 350°F, then the tools filtering this incoming data can cross-check against the location (latitude and longitude) and the time. And if the location is the same and the time is similar (say, 3:00 p.m.), then the incoming data is wrong. Big Data quality at the metadata level implies improved consistency across the data suite. Reference to metadata provides a common basis for data validation, standardization, enhancement, and transformations.

Quality of alternative data

Alternative data holds the promise of niche analytics but, at the same time, presents the biggest challenge in terms of quality. Alternative data is neither owned nor controlled by the users of that data. For example, discussion blogs, tweets, and likes on a social media page are all contributors to the Alternative data. These data can very easily comprise fake data, news,[13] and unsubstantiated information on social media platforms. Verifying their quality using traditional tools and techniques is an almost impossible task. Suggestions on handling the quality of alternative data include critically reading the material, checking the sources, and comparing with others. Automation and optimization of processes can include ML algorithms to take over some of the basic activities of verifying alternative data.

Sifting value from noise in Big Data

The available data can come in with a lot of noise that is irrelevant to the outcomes and does not gel with other data points within a data set. Analytics on this data will be embedded within business processes. Thus, the quality effort is in identifying the relevant data, providing it with some structure to make it analyzable, and then decision-making (explicit). Supporting data quality is the quality of the enterprise architecture, analytical algorithms, and visualizations. Quality initiative is an effort to sift value from the chatter and noise of data and make it available to business, verifying and validating its results through a business process. Techniques such as data sourcing and profiling, data standardization, matching and cleansing (scrubbing), and data enrichment (plugging the gaps and correcting the errors) are applied here to make the data analyzable.

Retiring the data safely and securely after use – data retirement after use (and when it is no longer current and relevant) needs to be undertaken carefully to ensure that the retired data cannot be abused by unauthorized parties. Besides, the data that is retired for one business can still have some potent value in it for another business – such as being able to track the history of decisions made by a business.

Quality in retiring data

A formal archival process of retired data is another important ingredient of quality. Spent or unusable data has to retire in a safe and secure manner. Audit trail needs to be preserved in order to provide the proof of data destruction in a controlled manner. Big retirement of Big Data is controlled by governance and compliance requirements. Governance and policies around the management of risks help control the retirement and detection of data sets. Attention to the external sources of data and systems is also required. The policies and processes provide checks and balances in data handling – including its cleansing, storage, usage, and eventually retirement. Unique data types such as meta- and alternative data may not belong to the organization. Hence, these data types may not retire. This archival process is undertaken with the help of data manipulation tools.

Velocity testing

Velocity of data at the entry is an important quality challenge requiring the use of tools for filtering. This is particularly so when the data is being generated through the Internet of Things (IoT) and machine sensors. The effect of bad data on quality is exacerbated if that data is generated by sensors and, as a result, is inundating the entire analytical systems.

Testing the velocity of data is important because of the impact of velocity on performance. This is the performance of the storage systems, as well as that of the analytics to keep up the processing with the velocity.

The following are the characteristics of velocity testing:

- Velocity testing includes testing the speed with which data is being produced and received (e.g., data generated by IoT devices).
- The rate of change of data (including speed of transmission, storage, and retrieval) and its impact on the analytics is also tested here.
- Also included is testing the speed of analytics – that is, the rate of processing of data and the creation of information or knowledge.
- Velocity testing will require the creation of a test environment that mirrors the production environment, as this is a part of operational testing of the performance of the system.

Visualizations are presentations on various user devices. These are the graphic user interfaces presenting the dashboard of analytics. The quality of visualizations plays an important role in providing a satisfactory user experience.

GOVERNANCE–RISK–COMPLIANCE AND DATA QUALITY

Effective governance is based on a body that comprises both business and technical decision-makers of the organization. Underneath this group of decision-makers is the business capability competency group that helps align the capabilities of the organization to the business outcomes.[14] These groups synergize operational, strategy, and IT professionals to ensure that the relevant IT systems, services, and platforms support the organization's outcomes.[15]

The GRC returns a significant value when it is carefully mapped to business capabilities. This is so because apart from ensuring compliance, GRC is also geared to ensure an ROI for the business. GRC ensures that BO adoption is of value to the organization.

GRC is helpful in maintaining compliance with both external and internal legal, audit, and accounting requirements. GRC coupled with business capabilities is vital to pave the path for AI technology investments.

Business compliance and quality

Business compliance is the need for the business to develop capabilities to meet regulatory compliances. These compliances enhance quality and reduce risks. The external demands for government and regulatory requirements also need the business to reorganize itself internally. An Agile internal business structure is able to respond better to ever-changing legislation. Consider, for example, the Sarbanes–Oxley (SOX) legislation. This legislation provides protection from fraudulent practices to shareholders and the general public and, at the same time, also pins the responsibility for internal controls and financial reporting on the chief executive officer (CEO) and the chief financial officer (CFO) of the company.

AI-enabled Agile business carries out this accountability and responsibility through changes in the internal processes, updating of ICT-based systems to enable accurate collection and timely reporting of business data, and changes in the attitude and practices of senior management. Another example of the need for the business to comply is the rapid implementation of regulations related to carbon emissions. This legislation requires businesses to update and implement their carbon collection procedures, analysis, control, audit, and internal and external reporting.

GRC complements the BO initiatives in an organization. GRC provides a consolidated and comprehensive approach to controlling an organization's business. GRC helps control existing enterprise data and functionality as much as it helps in handling the new Big Data. With a formal GRC in place, an organization can monitor its activities, provide necessary controls around the activities, conduct audits, and prepare reports. As a result of GRC, an organization improves its ability to prevent fraud by providing transparency and enabling executive-level control of data and business processes. This makes it imperative to discuss GRC in terms of the quality and value.

GRC, Business, and Big Data Governance within a business imply the following:

Governance is the overall management approach to control and direct the activities of an organization. This direction in turn requires an understanding of the desired business outcomes, and the capabilities[16] of the advent of Big Data require even more governance than before because of the uncertainty of data sources and the collaborations required among business partners.

Risk management supports governance through which management identifies, analyzes, and where necessary, responds appropriately to risks. Risks in the Big Data age are extremely dynamic because of the dynamicity of the underlying data and the depth of analytics. While Big Data analytics can help identify risks, there are also risks associated with the analytics themselves. The need to test the algorithm for syntax, semantics, and aesthetics, as well as for performance and other nonfunctional parameters, is acute in the Big Data world.

Compliance means conforming to stated requirements, standards, and regulations both external and internal to the organization. Due to the complexity of Big Data, compliance assumes greater importance than before. This is because compliance requirements (especially external and legal) can be potentially broken at any of the layers of the organization at which Big Data analytics are aiding decision-making. Analytics enable decision-making at the lowest rung of the organization, but it is the senior-most directors of the company that are responsible for the ultimate outcome.

Quality of service

Analytics are "served" through various services that are predefined. Quality of these services is assured by GRC. For example, the ITIL standards ensure areas of service management. Big Data services require the management of requests. The request management processes are modeled, reviewed, and tested to support a new service, process, and operations. The ability of service in the organization considers the following:

- Accounts and permissions – Request for new accounts, closed accounts, and permission changes.
- Projects – New development to be managed as a project. Development that requires complex management, multiple stakeholder engagement, and taking typically more than five business days of work. May have own release cycle or be released as part of other releases.
- Enhancements – Additions, extensions, and enhancements that can be done in typically less than five business days. This work is clearly defined, easily accomplished, and simple, testing overhead. Enhancements are mainly released as part of a planned cycle but may be released out of cycle.
- Defects – May be remediated as part of incident management or within problem management. Defects may take more than five days to remediate and may require project management. Defects are mainly released as part of a planned cycle but may be released out of cycle. Defects are mainly managed as incidents. For a request to be processed, the service will need actionable (all required information) and authorized (from the correct party) requests, especially when working with vendors and outsourced operations.
- Metrics and measurements associated with Big Data can provide performance information, risks, and opportunities for operational optimization. These metrics can be applied to measure analytics and management.

CONSOLIDATION WORKSHOP

1. What are the aspects of quality and their application in a digital business?
2. What are quality, security, and privacy functions in a data-driven organization?
3. Why is the quality of data the key to data-driven decision-making?
4. Describe the various aspects of quality in Big Data solutions?
5. What is the relationship between functional and nonfunctional (operational) aspects of quality and testing in Big Data management and analytics?
6. Why is it important to verify the syntax, semantics, and aesthetic quality of analytics, technologies, and visualization?
7. How are contemporary testing approaches applied with agility and continuous testing, to Big Data?
8. What is the importance of conducting normal, stress, and volume testing as part of the nonfunctional (operational) testing of Big Data solutions?
9. Describe governance, risk, and compliance (GRC) in the context of Big Data.
10. What are some discussion points around the quality of service and support using a governance framework?
11. What are the risks of bad data from getting into the decision-making process?
12. What are the differences between Big Data quality as compared with traditional data quality?
13. What are the quality assurance and control functions in ML algorithms?
14. How are data and algorithms audited?
15. Why is traceability of data in business processes important?
16. What are the policies and practices for individuals, businesses, government, and society?
17. Explain GDPR and other legislations (or lack thereof).
18. What are some considerations when sourcing third-party and compliance (GRC, government, etc.) data in optimizing business processes (leasing, purchasing, and other forms of sourcing including crowd-sourcing of Big Data)?

NOTES

1. Sherringham, K., & Unhelkar, B. (2010). Achieving business benefits by implementing enterprise risk management. *Cutter Consortium Enterprise Risk Management & Governance Executive Report*, Vol. 7, No. 3. Boston: Cutter.

2. In further discussions with Asim Chauhan, CEO, at www.riskwatch.com.
3. https://www.govinfo.gov/content/pkg/COMPS-1883/pdf/COMPS-1883.pdf, accessed 14 Oct 2020.
4. HIPAA regulation - https://www.hhs.gov/hipaa/for-professionals/privacy/laws-regulations/combined-regulation-text/index.html, accessed 14 Oct 2020.
5. https://gdpr.eu/tag/gdpr/, accessed 12 Oct, 2020.
6. Chen, C., Zhao, X., & Stamm, M. C. Generative adversarial attacks against deep-learning-based camera model identification. *IEEE Transactions on Information Forensics and Security,* Doi: 10.1109/TIFS.2019.2945198.
7. Goodfellow, I. J., Shlens, J., & Szegedy, C. (2014). Explaining and harnessing adversarial examples. *arXiv preprint arXiv:1412.6572.*
8. Unhelkar, B., & Sharma, A., (2017, March), Innovating with IoT, Big Data, and the cloud. *Cutter Business Technology Journal,* Vol. 30, No. 3, pp 28–33.
9. Unhelkar, B. (2005). *Verification and Validation for Quality of UML Models.* Hoboken, NJ: John Wiley & Sons.
10. Unhelkar, B. (2018). *Software Engineering with UML.* Boca Raton, FL: CRC Press, Taylor & Francis Group /an Auerbach Book). Authored, Foreword Scott Ambler. ISBN 978-1-138-29743-2.
11. Unhelkar, B. (2003). *Process Quality Assurance for UML-Based Projects.* Boston, MA: Addison-Wesley.
12. http://www.bpmn.org/ part of OMG (Object Management Group), accessed 14 Oct 2020.
13. https://www.mindtools.com/pages/article/fake-news.htm has interesting take on how to spot fake news, accessed 14 Oct 2020.
14. See note 1.
15. Tiwary, A., & Unhelkar, B. (2017). *Outcome Driven Business Architecture.* Boca Raton, FL: CRC Press.
16. Tiwary, A., & Unhelkar, B. Enhancing the governance, risks and control (GRC) framework with business capabilities to enable strategic technology investments. *Presented at the Proceedings of SDPS 2015 (Society for Design and Process Science) Conference,* Dallas, TX, Nov 1–5, 2015.

Chapter 9

Cybersecurity in BO

Significance and challenges for digital business

CYBERSECURITY ASPECTS IN BO

Cybersecurity is integral to digital business. Cybersecurity in this discussion is considered as a part of the overall Business Optimization (BO). Cybersecurity is a differentiator for customer value, protecting not only data but also the company perception and customer trust. This is so because the cybersecurity image of the company impacts customer value. Cybersecurity vulnerability in the organization's systems can lose customers and subject the organization to litigations. Proper handling of cybersecurity protects customers and revenue, confidence, and accelerates growth opportunities for the business. A holistic approach to cybersecurity has a positive impact on the organization's ability to handle disruptions, increase resilience, ensure compliance, and enable business continuity. Cybersecurity discussions in this chapter complement GRC and quality topics discussed in the previous chapter.

Cybersecurity is considered across a broad range of AI applications in BO. Starting from the edges of the organization (customer touchpoints), through to analytics, user behavior, and the Cloud, cybersecurity applies everywhere and across the entire organization. The four aspects of *Think Data* discussed in Chapter 2 (Figure 2.1) are revisited here from the cybersecurity perspective. Security of the "handset," "dataset," "toolset," and "mindset" each requires an understanding of devices, analytics, processes, and people. Cybersecurity standards and framework help protect data and enhance customer value. Timely communication is another crucial part of cybersecurity as a business strategy. Leaders, managers, network admin istrators, users, and customers form an important part of the communication strategy.

A digital-savvy business using data in decision-making assumes responsibility for the safety of that data. Cybersecurity complements GRC and quality initiatives, which were discussed in the previous chapter. This chapter discusses the crucial aspects of cybersecurity in BO with the goal of ensuring customer value.

Cybersecurity functions

Cybersecurity in BO has two key functions: securing the optimized business processes and using AI to undertake cybersecurity intelligence (CI). Cybersecurity is thus securing the data and business processes, and the use of data and processes in detecting and preventing breaches. Descriptive, predictive, and prescriptive analytics are used to detect and prevent security breaches. Data analytics assumes importance in developing cybersecurity strategies.

The typical data life cycle starts by ingestion through devices and users. The data then travels over the networks. The cloud is the common mechanism to store and analyze data. Eventually, the results are presented via the networks onto the user devices for decision-making. Cybersecurity is required at each point in this data journey. Furthermore, cybersecurity also ensures that the data is retired correctly.

The cybersecurity function is only limited to a real threat or breach. At times, the *perceived* threat is as important as a real threat and requires similar, substantial handling. This is so because the perception of security impacts decisions by customers and partners. Therefore, the perception of security of an organization's data is as important as, say, its regulatory compliance of security and privacy.

AI in the cybersecurity space is a double-edged sword. Attackers tend to use the same AI technologies and analytics that defenders use. For example, attackers hide potentially malicious scripts by encoding in the same way defenders do. The AI technology is unable to distinguish between the attackers and the defenders. Attackers use AI to threaten, breach, and abuse businesses and people. The ethics and moral aspects of AI and security become equally important in the cybersecurity discussions as alluded to separately in this chapter.

Cybersecurity as a business decision

Cybersecurity breaches cost trillions of dollars to businesses globally. The cybersecurity risks are, however, far more significant and complex than the dollar figures indicate. Embedding AI within business processes increases their complexity and usage of data. The dollar measure is an insufficient way to measure the cybersecurity effort. Neither can cybersecurity be considered as a purely technology issue. While the data and analytics technologies are at the root of the cybersecurity challenge, the business context remains foremost in implementing cybersecurity within BO. Thus, cybersecurity capabilities are considered a function of people, process, and technology. This business context is a function of customer needs, business goals, security costs, and performance of business processes. Cybersecurity function thus assumes business responsibility and is a business decision more than a technology one in business optimization.

As a business decision, cybersecurity risks are balanced with the efficiency of business processes. For example, a completely and heavily secured business can suffer business performance attrition to a level where it loses customers. Therefore, the business mindset has to balance the risks of security with that of performance loss in order to arrive at the right level of security. This balancing act requires ongoing, agile management of the cybersecurity function.

A strategic approach to cybersecurity is proactive in nature. Such an approach comprises business goals, technical capabilities, availability of resources, and the business environment. Outcome-driven business architecture[1] helps in aligning cybersecurity capabilities with the desired business outcomes. The cybersecurity capabilities are further prioritized using a risk-based performance metrics. Cybersecurity decisions, based on performance analytics, are made visible to staff, customers, and regulators.

Cybersecurity strategies require the development of controls around the various data processes (refer back to Figure 2.3 wherein processes around data are discussed). Security standards provide the basis for adequate, reasonable, consistent, and effective cybersecurity controls. The controls should also credible, demonstrable, and auditable.

Cybersecurity and penalties

Risks emanating from the regulatory bodies are exacerbated in data-driven digital businesses due to the myriad rules and regulations surrounding the use of data. Risks due to lack of control can also incur substantial penalties. These penalties are business issues, because apart from the financial losses the business also stands to lose substantial goodwill resulting from penalties and litigations. Data-driven businesses can also fall prey to money laundering and other financial misdemeanors.

Financial risks are an important consideration in cybersecurity budgets. Understanding these varieties of risks, the corresponding value generation, and the investments needed in the cybersecurity space is also a balancing act. Stringent cybersecurity implementations increase costs and may degrade performance. Customer experience is an important factor in ascertaining security levels. Cybersecurity tools help in the early identification of threats and provide insights to handle their eventuation.

Cybersecurity challenges during BO

As AI is embedded in business processes, it changes the way the business operates. This change to business processes creates a potential for hacking especially as the newer business processes are data-driven, complex, and not always explainable. Data-driven business processes tend to be black boxes with a minimal explanation of what is inside. In such situations,

AI is seen with trepidation by both users and business leaders. The lack of explainability of AI can also insert security-related doubt and uncertainty in the minds of customers.

Cybersecurity function covers the security and privacy of data and metadata. For example, cybersecurity deals with the security of location-independent, cellular metadata that may not be owned by the company. In contrast, the cybersecurity function has to also protect the large amount of data within the firewall of a company. Additionally, technologies such as blockchains and ethereums that are becoming part of the digital business present security risks that cannot be handled by a single business on its own. These technologies require a collaborative, industry-based approach to security.

Outsourcing of AI development and use of third-party data can create a façade of shift of risks faced by the business during BO. Even though the service is sourced from a provider, the overall responsibility of cybersecurity is with the business providing the final service and value to the customer. Transfer of risks to a vendor or partner is a complex legal quagmire that stretches beyond technology and business decisions. Awareness of the challenge is required but dealing with the actual threat of such multiparty cyber risks is beyond the scope of this book.

Cybersecurity vulnerabilities and impact

Data-driven organizations accept and handle risks related to data and their application in processes. These risks make the organization vulnerable to cyberattacks – both external and internal. Existence of vulnerabilities is inevitable in digital business due to the vastness of data and complexities of algorithms. As a result, AI-based tools become indispensable in cybersecurity.

Furthermore, the impact of a breach is not always proportionate to the actual vulnerability. A small breach in the enterprise's systems, networks, storages, and processes lead to compromising the entire system. For example, Alex Hope[2] relates the story of hacking into former Australian prime minister Tony Abbott's by simply gleaning details from the boarding pass posted by the latter thanking the flight crew. The booking reference printed on the boarding pass provided access to the booking and a search through HTML code revealed the passport and phone details. Thus, an apparently innocuous action lead to a breach in the security.

Assumptions, opinions, and the slightest variation in the input data can skew the results generated by analytics. Attackers in cybersecurity space capitalize on each of the aforementioned aspects and present a threat. For example, a ransomware attack need not mess all records in the database, but only one record – and demand ransom to clean or unlock that one record which is not revealed to the owner business. "Fake news" is a formal term indicating this phenomenon of skewing the output and, thereby, perception with a minuscule variation in the input.

Tools, techniques, and systems are applied by businesses to secure their data and business processes. For example, multifactor authentications are now a norm rather than an exception. Configuration of networks and systems still play an important role in security, and poor handling of their operations can also present security challenges. Similarly, physical security of devices becomes a vulnerability of concern as it tends to get oft neglected while the focus remains on the software and data security.

Cyber attacker's psyche

Cybersecurity is as much a mind game as a data and technology one. As businesses aim to secure their optimized processes, attention is required to the psychological aspect of security as much as to the technical one.

To begin with, the psyche of every attacker is nefarious. The unethical mind of a cyberattacker may have many reasons – some even justified in their own minds (e.g., seemingly unfair dismissals from a company or desire to "settle scores"). Attackers find dedicated time and effort and work without being paid for it as compared with the typical defenders. Attackers aim for illegal financial gains but money is not the only purpose to threaten an organization. Attackers continue to grow their capabilities alarmingly and act proactively to hurt businesses. In comparison, defenders are usually reactive. Attackers can sidestep defender tools by identifying their characteristics and overcoming them. Attackers can encrypt and hide the information being sent the same way defenders do.[3] Attackers overcome encryption with tools that can help them bypass filters and evade detection. Cybersecurity strategies benefit with the help of AI to identify potential areas of threats, likelihood of breaches, and the impact of those breaches on business. Defenders can be provided with greater resources than are usually made available to them, including tools and skills, to thwart the attacks.

The psyche of an attacker is to masquerade as a legitimate user sending legitimate messages. The masked messages from attackers aim to infiltrate targets in any way possible. Examples of such behavior include text encoding, slipping past filters, and delaying defenders from figuring out what is going on. Attackers aim to hijack an account, spread web worms, access browser history and clipboard contents, control the browser remotely, and scan and exploit applications. Attackers also study the defender psyche and capitalize on lax user behavior and their possibly untrained mindset. For example, attackers start with a list of emails of the people they are targeting in an organization (or even a country). Initial attack attempts are made across the board to identify the "weak" users. Profiles of such users are then built in order to create a sustained attack.

As business processes are optimized, the analytics within those processes provide valuable decision-making insights. Reliability and trust in those analytics are absolutely essential for BO to succeed. Cyberattackers do not need to do much if this reliability and trust are broken. Therefore, cybersecurity strategies need to incorporate this crucial understanding of the attacker's psyche in developing multipronged approaches to cyber defense of the data, analytics, processes, devices, and users.

SECURING THE OPTIMIZED BUSINESS

Developing cybersecurity strategies requires a technical understanding of the types of cyber threats. These cyber threats pose differing challenges to the optimization effort. Threats are external as well as internal. Business analysis capabilities aid in understanding the types of cyber threats and developing strategies to counter them. The complexity of networks and data storages in BO is such that there is no point in time where a cybersecurity strategy can be said to be complete. Cyber strategies are continuously evolving in an agile manner. Traditional intrusion and detection analytics are based on log files. A more strategic approach uses the latest techniques and tools for collecting and analyzing network traffic datasets as a "trade-off between expressiveness and speed."[4] This is so because cybersecurity as a business decision has to balance network transmissions, IoT devices, and Cloud searches with costs, performances, and people behaviors. The effect of a particular strategy is understood only after it is implemented. Therefore, agility in developing and maintaining cybersecurity is a must.

Types of cyber threats

Cybersecurity data has a range of characteristics that are important in the understanding of cyber threats and developing countermeasures. For example, even if an organization has collected a large amount of data related to security, it is important to know how current is the data, how relevant it is to the organization, and what is the impact of a threat or a breach on the business. At times, a seemingly small breach can have a large impact on the reputation of the business. Cybersecurity analytics assist in developing an understanding of the availability of resources, possibility of early detection, readiness of response, and approach to recovery.

Figure 9.1 summarizes the use of analytics and analysis in the cybersecurity space.

Figure 9.1 also summarizes four types of cyber threats: malware, phishing, eavesdropping, and denial of sources. Additionally, these threats can materialize from outside the organization or from within. A brief description of these cyber threats follows.

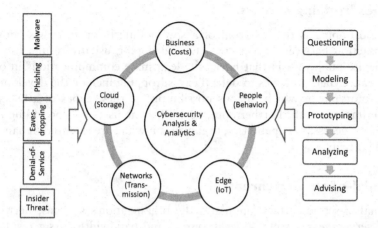

Figure 9.1 Using cybersecurity analysis and analytics in securing business optimization.

Malware threats

Malware is malicious software that gets embedded with business processes. Malware breaches a network through a vulnerability like a user downloading or clicking a link embedded in an email or email attachment. Automation and optimization can inadvertently introduce malware within business processes. This can result in the installation of additional viruses that will proliferate on their own, spy on activities, and block access for legitimate users. Ransomwares encrypt in particular the data thereby making it unusable. Encryption key of such malwares is known only to the attacker and Cryptocurrency (Bitcoin) payments are demanded. Spywares email attachment, ransomware, viruses, and worms to the user. Cybersecurity strategies include scanning of all incoming messages in various formats for malware bots. Anticipating bad messages with encryption that are masquerading as good messages is essential for defense. A database of all known malwares and ransomwares as well as forming a community of users with similar understanding is another approach to handle malware.

Phishing threats

Phishing appears to come in an email or phone message from a known or apparently reputable source. The message, however, is a fraudulent communication that is searching for vulnerabilities. Phishing searches and asks for sensitive data like credit card and login information. Cyber strategies dealing with phishing threats need to upskill the users on an almost daily basis. This upskilling of users includes the demonstration of the phishing messages and their impact on business processes.

Eavesdropping threats

Eavesdropping is the insertion of a spy module in an otherwise normal transaction between two parties. Eavesdropping usually results from an unsecure public Wi-Fi that hacks the legitimate communications in order to steal. Cyber strategies to defend eavesdropping include the development of prototypes in which various possibilities of these types of attacks are experimented with. Furthermore, upskilling the users and providing them with the necessary capabilities, organization wide, are essential in handling these threats.

Denial-of-service threats

Denial-of-service attack inundates the organization's systems, networks, and servers, exhausting all its resources and bandwidths. As a result, the business systems cannot fulfill legitimate requests. Multiple devices that have been compromised are used to launch the DOS attack. Cybersecurity strategy must focus on the sensitivity of the service, creation of mirrored service capabilities, and continuous scanning of the systems for their performance. Slightest hint at a drop in service levels is investigated using automated flags and the possibilities of denial of service obviated.

Insider threats

Cyber threats are usually considered coming from external parties. While this is true in greater percentages, at times the employees or confidants of the business can also attack the organization for reasons different to the external attackers. These insider threats result from disgruntled employees, whistleblowers, and spies. Insiders can steal data, credentials, and data storage or simply exfiltrate large amounts of data.

Tracking user behavior with analytics can help narrow down the possibility of insider threats. The risk in such analytics is the potential biases of the analytics. Cybersecurity implementations need to collect data and logistics keeping sensitivity of the insider threat in mind. For example, more surveillance of a person can tarnish their image even if they are clean. People may feel marginalized by the surveying once the surveillance becomes known. Protecting physical data and resources from insiders requires the use of multiple verifications from multiple sources, strict inventory control, and NI-enabled data analytics.

DEVELOPING CYBERSECURITY STRATEGIES

The cybersecurity strategies provide the roadmap for implementing security measures. Developing cybersecurity strategies includes data, analytics, and metrics. Cybersecurity data is sourced from various external providers in addition to being collected internally. Partnering businesses, clients, and vendors add to this large collection of security data. Regulators and policymakers make security data more widely available for use in order to boost the security of the entire business community. Data is extracted, analyzed, and visualized when the security strategy is implemented. Cybersecurity analytics use a combination of the data generated and owned by the organization along with third-party provided security data.

An important aspect of data-driven cybersecurity strategy is metrics and measures. Metrics can help understand the level of security and vulnerabilities of handset, dataset, toolset, and mindset. Each aspect of the aforementioned *think data* needs a metrics and measurement program around it.

Cybersecurity strategies also need to incorporate the visualization aspect of data. Data drives cybersecurity analytics, but it must be handled appropriately. Visualization of analytics has to be intuitive, relevant, and understandable in discovering vulnerabilities. Visualization should also not jeopardize the security of what is being presented.

The argument for using data-driven analytics and analysis to make security decisions is based on security, data collection, and eventual insights. Data analytics reduce the impact of a data breach based on vulnerabilities. Vulnerabilities expose valuable data to bad actors. Data-driven security practices record known exploits in successful data breaches and analyze them to spot future vulnerabilities. Correcting vulnerability reduces the probability of the impact of data breaches. Vulnerabilities do not account for all common exploits. Therefore, vulnerabilities are identified and prioritized based on their high-impact. Attackers focus on the rare vulnerabilities than the well-known ones. Therefore, vulnerabilities with less-known exploits need to be prioritized higher as they are likely to be attacked first. AI-based data analytics can provide insights for prioritization. Furthermore, dedicated resources and fine granular analytics enable a balance between known and unknown exploits.

Organizing cybersecurity data and functions

Cybersecurity data and functions are organized in a multilayered format. These layers apply to the data, processes, and people. Supporting these analytics are tools that offer antivirus protection and file verifications. At the device level, data is collected by placing sensors in the right domain and with the right vantage. At people level, training and educating the users is also a multilayered activity.

At the core technology level, multilayered spam filtering creates multilayered data that is subject to analytics. This data is collected and filtered for activities. The suspicious data is analyzed. Comparison of data and data

Table 9.1 Organizing the cybersecurity function at a technical level

Aspects	Description
Domain	Network, Service, Host Active domains form the broad description for security data collection.
Vantage	Location of sensor packet is determined in vantage by interaction between the sensor's placement and the routing infrastructure of the network.
Action	What does a sensor do with the data? For example, report it, initiate a response, control the analytics, and preserve them.
Validity	Strength of the premise of analytics based on sensor data collected. This validity also depends on currency of data – what is valid just now may not be valid after an hour or a day.

sources to a list of known risks is undertaken. Spammers of data collection and their domains as a likely source of spam or malware are also part of data collection.

Table 9.1 summarizes the organization of the security function. The four aspects of these security functions provide the framework for positions of sensors and collection of data.

Security data source is part of security analysis. This security data is gained by multiple sensors throughout a network. The sensors capture source data from different parts of the network and store it into readable files. Sensors are placed as network taps or they appear in firewall logs. Collecting a large amount of data from any network is not the goal. Quality security data is obtained by the density of security information with minimal overload. Removing redundancies from logs, increasing reliability of information of the connections between IP addresses, and validating domain names are some examples. Vantage is another important aspect of data collection. Source destination of data is improved by proper vantage. For example, data coming into a router that is being split to a single workstation and also a switch that is going to a different workstation as well as a vantage ensure data is accurate and knows the exact place of its origination.

Cybersecurity data collection has to consider the devices used by the end-user and those used for data collection as both are changing continuously. These devices (sensors) and their positioning in organizational networks are part of data collection. Cybersecurity data collection also identifies the formats packets and filters used. The speed and accuracy of data transmission as well as breach detection, are factors of network architecture that limit the data captured and filtered from each packet.

Cybersecurity data analytics

Figure 9.2 shows the Agile (iterative) approach to handling threats in a digital business. The iterations in the inner and outer circles in Figure 9.2

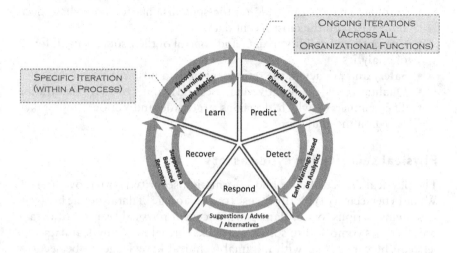

Figure 9.2 Agile iterations in detecting cybersecurity threats and response/recovery actions.

represent the immediate and the ongoing series of cybersecurity activities and actors.

Cybersecurity data is explored and analyzed through a combination of descriptive-predictive analytics. These analytics are carried out iteratively in order to narrow down on interesting possibilities of threats and breaches. Typically, cybersecurity analytics start with understanding the contents of the logfiles. Further, filtered data enables analytics on source and destination IP and produces outliers or skewing towards suspicious or abnormal pattern of deviation from standard expectations.

Univariate descriptive analytics (e.g., histograms, bar charts, boxplots) are used to analyze the logfiles for quantitative security variables. Bivariate (e.g., scatterplots) and multivariate analytics can also be applied depending on the sensitivity, urgency, and application of the analytics to cybersecurity. Data analytics tools collect, analyze, and visualize various security insights.

The design of analytics on security-related data has to keep the following considerations in mind:

- Architecture of the sensor network and the positioning of sensors across the network – including wired and wireless sensor networks
- Existing repository of cybersecurity data and the frequency of updates to that data from the sensors in the network
- Reliability and relevance of processing of the data and the use of analytical insights in security decisions
- Granularity of analytics – as coarse granular analytics will work typically on historical, large datasets to identify patterns whereas fine

granular analytics provide on-the-spot insights for immediate decisions based on the most recent data
- Currency and validity of data and control of their sources in all forms of analytics
- Safety and privacy in the retirement of data post-analytics
- Qualitative and quantitative analytics and manner of their visualization
- The manner in which decisions are made and actions undertaken based on the analytics

Physical security for cyber assets

The physical aspect of data and information security is often overlooked. While cybersecurity typically focuses on the storage of data and the breaches in systems, serious loss or damage can occur if physical security of data is breached. As compared to software attacks, breaches of physical data centers can be carried out with minimal technical knowledge. Cybersecurity strategies ensure physical data centers are secured against physical attacks, accidents, or natural disasters. Physical locks, biometric access, manual security, surveillance with cameras, intrusion detection, and auto reporting are important aspects of physical security.

GDPR and Industry Data Security Standards also require organizations to monitor access and restrict unauthorized entry in any facility that stores, processes, or transmits customer data. Similarly, HIPAA prescribes physical measures, policies, and procedures to protect a patient's electronic patient record. The measures dictated by these standards are incorporated in a good cybersecurity strategy.

Cybersecurity analysis using business analysis capabilities

Cybersecurity analysis (as against analytics) is the application of Business Analysis (BA discussed in detail in Chapter 6) capabilities to the security function of the organization. BA views and models the business processes holistically. Therefore, analysis plays an important role in developing a cybersecurity strategy than the mere statistical analytics of security data. Business analysis capabilities view roles, activities, deliverables, and supporting technologies simultaneously from the user's perspective.

BA activities include asking the right questions about security to the right people, documenting and modeling the answers and verifying the security of the processes. BA effort in securing the organization is a continuous, ongoing one that does not finish at any point in time. Therefore, BA capabilities and approaches are ideal to develop a holistic defense strategy. This continuously iterating approach to cyber defense is ideal for many threats that are unknown and remain so till the attack happens. Unknown vulnerabilities are a challenge to cybersecurity especially in the AI world where the vastness of data and the complexity of algorithms make it impossible for a manual, piecemeal approach to succeed.

Analysis makes use of data and analytics to create opportunities for security strategies in a proactive way. Data analytics includes descriptive, predictive, and prescriptive analytics on security data. Data, and in particular Big Data, provides opportunities to describe and predict security threats and occurrences of breaches. Examples of security data include networks, log-files, user behavior, and device locations. Security analysis together with data analytics identifies vulnerabilities in the optimized business processes.

Cybersecurity analysis, however, makes provisions to incorporate values and judgments in security-related decisions. For example, instead of relying entirely on AI to point to an attacker, analysis will also explore the possibilities of biases. Biases are subjective elements that can appear in the data (due to previous erroneous entry), algorithms (errors of misunderstanding), and decisions (human biases).

Cybersecurity standards and frameworks

The most popular cybersecurity standard is the NIST cybersecurity framework and ISO 2700x.[5] Another standard for Cybersecurity Readiness and Investment is the CARE standard.[6] These standards provide the background in developing cybersecurity strategies and aim to reduce cybersecurity risks for businesses. Standards include collections of tools, policies, security concepts, security safeguards, guidelines, risk management approaches, actions, training, best-practice assurance, and technologies.

Best practices and techniques in determining cybersecurity capability levels provide value to organizations implementing the standards and models. Desired levels of cybersecurity form the basis for prioritization. Maturity models measure how good the capabilities are. These matured capabilities are aligned to the goals. Cybersecurity capabilities need continuous alignment to organization context.

CARVER (Criticality, Accessibility, Recuperability, Vulnerability, Effect, and Recognizability)[7] provides yet another helpful framework in developing the cybersecurity strategies. Having originated from the US defense, this standard provides an approach to identify and rank specific targets so that attack resources can be efficiently used. As a result, CARVER can be used by digital businesses from an offensive (what to attack) or defensive (what to protect) perspective.

CYBERSECURITY INTELLIGENCE (CI)

Cybersecurity intelligence (CI) is the use of AI to enhance cybersecurity functions. Cybersecurity intelligence starts by identifying the security goals and acceptable risk levels. Prioritization of risks can be undertaken with standards and frameworks mentioned above. Cybersecurity intelligence starts with data, but the analytical part needs well-defined metrics and measurements. People, processes, technologies, and money are used in a balanced and holistic manner with the help of standards, metrics, and measurements. An understanding of data sources used is a valuable input in developing the balanced strategy. Due consideration to people issues is also helpful in approaching cybersecurity in balance. For example, human error, apathy, or negligence is the cause for more than half of security breaches. Most software viruses are activated by a click of the user.[8]

AI analytics embedded in business processes become extremely complex. AI-based cybersecurity tools become necessary to handle this complexity. Security metrics and measurements from the basis of CI tools design and choice of metrics, use of relevant analytical techniques, and understanding of the mindset provide the backdrop for the CI functions.

Change to a security mindset is a people issue and CI aids and supports it by narrowing areas for attention and action. Training and mentoring play a role in enhancing the cybersecurity mindset.

Cybersecurity metrics and measurements in CI

CI uses security-related data in order to understand vulnerabilities. The data is downloaded and analyzed using tools. CI is designed keeping the business outcome in mind and the use of cybersecurity tools. Cybersecurity metrics comprise external events (e.g., number of breaches) and internal responses (e.g., speed of detection). The number of threats and the occurrence of breaches are external and not controlled by the business, whereas the speed of detection is an example of the internal preparedness of the organization. Example metrics include total malware incidents, percentage attacks blocked, and so on. CI learns from every previous attempt to attack. And each learning provides the opportunity to imagine further attacks. CI helps narrow the areas of defense and make it cost effective. CI is part of the overall business strategy.

Breaches impact business processes and corresponding customer sentiment. Data indicates the strength of firewalls and filtering mechanisms. CI makes use of this data to measure the overall security design, data collection, transmission, and analytics. CI needs focus, position of the organization, and areas of focus. Detecting breaches and preventing further attacks have to be accompanied by transparency and honesty.

Customer value remains the focus of BO and, therefore, also of CI. Since the perception of security by the customer is as important as actual security, CI includes analysis of customer sentiments. For example, if customers are unable to easily gain access to services, then even for a highly secured business, the perception and quality of the offerings suffer. Data analytics not only prevent and detect breaches but also provide insights into customer perceptions of security.

CI is both strategic and tactical. Examples of tactical CI include analytics on the daily/hourly number of breaches. The ease of user login or lack thereof is also tactical to CI. CI at a strategic level provides the overall security position of the organization. Tools, technologies, systems compliance audits, and competitor positions are of concern in strategic CI. Organizing training in security to update the mindset of employees is crucial and strategic.

CI includes a detailed plan of action in case of data security breaches and related business disruptions. Actions are based on an understanding of the business and competitors. CI aims to learn not only from the data within the organization but also from across the industry.

Levensthein distance as a measure in CI

An example of CI is the application of analytics to cybersecurity data such as the user logfiles. The contents of the logfiles include log-ins, frequencies of log-ins, and relevance of users. Simple length data analysis is based on header-length and record-length. Median, average, and standard deviation are derivatives of value. Earlier filters can be used for further analytics for nonurgent, nonstandard, and unexpected flags. Accessing and manipulating the log files are cybersecurity analytics.

An example of cybersecurity defense technique is the calculation of the Levenshtein distance – which is a "string distance" between phishing domains and authentic ones. For example, the Levenshtein distance between "hello" and "hallo" is 1 – which quantifies the difference between phishing and authentic domains. Defenders can use AI to quantify this distance for a large number of domains to start flagging phishing emails. A phishing attack attempting to redirect the user to "c0rp.adomain.com" instead of "corp.adomain.com" can be caught this way.

Base rate fallacy in cybersecurity measure and the validity of positives and negatives in CI

CI continually deals with sensitivity and specificity of potential threats and breaches. Security-related data is collected through sensors which are not only devices but also any other binary classifier.[9] Collection of such cybersecurity data is followed by its analytics. Each sensor data has potentially false positives and false negatives. CI analytics requires a balancing act (or tradeoff) between the potential "false" readings.

This balancing act is based on the "base rate fallacy" that gives insight into the use of CI. For example, a 99% accuracy can produce 100 false positives for a 10,000 sample. These false positives are not equally spread across the sample. Therefore, these tests cannot be entirely used for detection on a network.

Intrusion Detection System (IDS) has to consider multiple types of probabilities based on false positives and false negatives. Bayes'[10] theorem gives the probability of an event and is used in CI. Considering security data in a mono-dimensional leads to what is called Base Rate Fallacy.

An IDS can incorporate the concept of Base Rate Fallacy[11] in its algorithms. Consider, for example, face recognition as an identification. Face ID also works on percentages. For example, if an ID is 99% accurate for face detection, it still leaves many opportunities for a face to gain illegal to entry. IDS can factor in the possibilities of fallacies in its algorithms.

Filtering algorithms for email phishing for CI

CI also manages email rules and filters phishing emails. Managing emails and filtering requires the application of AI. Analytics on emails establishes a pattern for filtering. While the filters filter out suspicious emails to the spam folder, the rules for filtering can create an imbalance. Filtering algorithms model the email route, available hardware, networks, and devices. Regular internal monitoring of emails and daily updates is an essential element of CI.

Transmission of emails, especially on the cloud, each requires a security focus. Emails on the mail server can be accessed anywhere depending on how the administrator sets it up. These mail servers are susceptible to attacks. A service gateway is needed to create the email rules and filtering. User location and access is out of the administrator control.

CI scripts parse through large datasets to find a string, manipulate delimiters, split them by standards and rules, and arrange them in specific patterns of possible encryption types. Once the patterns are established, techniques for text analysis to decrypt the data are used. The techniques produce the results that develop an understanding of how the encryption works and fine tune another script specific to that encryption.

AI-based approaches to cybersecurity scale up to counter the cybersecurity risks. Cyberattackers also use tools, typically bots, which continue to ping for vulnerabilities. Bots form a network of their own to scale the breadth and depth of attacks.

Detecting cyberattacks requires the use of AI tools because of the need for speed and granularity. This role can be played by AI technology. ML algorithms can gather past security including attacks and breaches. ML algorithms are then trained on such data to detect anomalies.

AI has its own vulnerability. For example, changing an image in a way imperceptible to humans can mislead an AI program to label the image as something entirely different. Alternatively, it is easy to produce images that are completely unrecognizable to humans, but that state-of-the-art AI programs believe to be recognizable objects with 99.99% confidence.[12]

Tools for cybersecurity intelligence

Cybersecurity analytics need tool support. SIEM tools provide this support. The security factors considered by SIEM include: ports and protocols, size, IP addresses, time, TCP options, helper options, and filtering options. SiLK[13] is a standardized SIEM used by many organizations. Other examples of tools include Splunk and Webroot; SPAM Titan and Barracuda are typically used in email filtering and analytics. These tools ingest datafiles, suitably format them and manipulate fields in order to understand security breaches and create security alerts. SIEM correlates a lot of data information that aid cybersecurity.

The SiLK suite, also known as the System for Internet Level Knowledge suite, provides tools for the collection, storage, and analysis of net flow data. NetFlow is a router feature that enables the ability to collect IP network traffic.[14] SiLK provides a number of useful tools that can query NetFlow, activity, policy violations, and time frame of a particular event and analyze them in an efficient system. The security data is subject to quantitative and qualitative descriptive analytics. Quantitative data produces statistics such as average and standard deviation whereas qualitative data produces frequency and observations. This kind of information can be crucial with IP data when visualizing what source contacted what destination how often, for a basic example. The analytics are represented in numerical or statistical fashion (quantitative), or in the form of ideas or graphs (qualitative).

For example, the frequency of users accessing information compared to all users provides qualitative analysis. The total number of times a group of users is accessing the information asset per day, week, or month is a quantitative analysis. A query on the net flow files, including the IP addresses of users and specified time frame, provides a profile of access and usage.

CONSOLIDATION WORKSHOP

1. Why is cybersecurity business decision more than a technical one? Discuss in the context of cybersecurity as being a survival issue for digital business?
2. Discuss the cybersecurity challenges in BO under the *think data* umbrella.
3. Cybersecurity has a positive impact on the quality of data. Explain with an example.
4. Cybersecurity is a perception issue as much as a real one. Discuss with examples.
5. Discuss how cybersecurity of data across the Cloud is based on network security, storage security and retirement of data security. Provide examples.

6. Why is identification and prevention of bad and dark data challenging with Big Data?
7. Cybersecurity analytics is descriptive, predictive, and prescriptive. Discuss with examples. Also describe the difference between cybersecurity analytics and analysis.
8. What is malware? Phishing? Eavesdropping? Denial of service? Discuss with examples.
9. How is an insider threat different from an outsider one? What additional caveats does the organization need to thwart insider threats?
10. What are the roles of architectures in networks and predictability of breaches? How can network architecture be used in positioning of sensors?
11. Why is cybersecurity in collaborative business processes more challenging than business processes of a single organization?
12. Cybersecurity with efficiency of business processes is an important balancing act for performance. Explain with examples.
13. Describe how physical security is incorporated in cybersecurity strategies? Also outline how it can be implemented?
14. What are the four important considerations in positioning sensors (hint: Table 9.1)?
15. What is the role of training and mentoring in awareness of staff for security issues?
16. How can AI protect itself from malicious attacks?
17. What are the popular cybersecurity standards?
18. How are filtering algorithms designed and implemented?
19. Discuss the tools for cybersecurity. In particular, visit the SiLK site to understand how a tool can be used in the cybersecurity of an organization

NOTES

1. Tiwary, A., and Unhelkar, B., (2018), *Outcome Driven Business Architecture,* CRC Press, (Taylor & Francis Group /an Auerbach Book), Boca Raton, FL.
2. https://www.bbc.com/news/world-australia-54193764; accessed 4 Oct 2020.
3. https://www.base64decode.org/Links to an external site.
4. Collins, M., (2017), Network security through data analysis: From data to action. ProQuest Ebook.
5. NISTcybersecurityframeworkhttps://www.nist.gov/cyberframework;accessed28 Oct 2020; https://www.iso.org/isoiec-27001-information-security.html; accessed 28 Oct 2020.
6. The CARE Standard for Cyber security Readiness and Investment.
7. https://en.wikipedia.org/wiki/CARVER_matrix; accessed 5 Oct 2020.
8. https://www.cybersecurityintelligence.com/blog/cyber-security-awareness-month-october-2020-5244.html.

9. See note 4.
10. Bayes Theorem - https://plato.stanford.edu/entries/bayes-theorem/; accessed 28 Oct 2020.
11. Base Rate Fallacy - https://www.oxfordreference.com/view/10.1093/oi/authority. 20110803095449924; accessed 28 Oct 2020.
12. Nguyen, A., Yosinski, J., and Clune, J., (2015), Deep neural networks are easily fooled: High confidence predictions for unrecognizable images, *Proceedings of the IEEE Conference on Computer Vision and Pattern Recognition (CVPR)*, pp. 427–436.
13. SiLK – SEI of CMU - https://www.sei.cmu.edu/news-events/news/article. cfm?assetid=521324; accessed 28 Oct 2020.
14. See note 4.

Chapter 10

Natural intelligence and social aspects of AI-based decisions

THE "ARTIFICIAL" IN AI

Artificial intelligence (AI) imitates natural intelligence (NI). Therefore, it is called *artificial* intelligence as it is not the same as NI. Humans are able to contextualize the decisions, apply ethics and morals, and able to *feel* the joy of successful outcomes. These subjective factors make a huge difference in the outcomes. Therefore, it is important to explore the human aspect of AI-based decision-making in optimizing business processes. This chapter explores the "soft" aspects of business optimization.

AI optimizes business processes and increases productivity but leads to social, ethical, and moral challenges in decision-making. The machine learning (ML) algorithms mimic and augment human thinking processes but they are unable to explain the *reasons* behind the insights generated. The lack of reasoning or explainability is an important consideration in AI adoption. Deep learning (DL), in particular, is multilayered and complex, making it impossible to ascertain the reasons behind the insights generated. Explanations, however, are important from both development and usage viewpoints. Philosophically, Agrawal et al.[1] ask "Is AI an existential threat to humanity itself?" This question is asked across business, social, political, and various other sectors. Ada Lovelace[2] clarified the impossibility of AI taking over humans entirely, decades ago. The discussion in this chapter underscores the importance of human inputs in decision-making because AI does not have the same reasoning, contextualization, and sensitivity as that in human decision-making.

In discussing AI impact beyond business, Agrawal et al.[3] talk about three tradeoffs: productivity versus distribution, innovation versus competition, and performance versus privacy. In each of these three tradeoffs, there is a substantial element of intelligence that is beyond AI. As a result, AI cannot take over issues associated with ethics and morality that transcend the algorithmic or legal frameworks. Neither can AI take over innovation – which remains in human purview. Indeed AI is used in supporting the decisions in a balanced manner.

AI supporting human decision-making is qualitatively different from AI taking over decision-making altogether. Whether it is a decision related to a credit or a diagnosis related to a disease or a job candidacy interview in a human resources department, AI without any human intervention can be potentially catastrophic. The extent to which humanization of optimized business processes should occur is, however, a subjective decision. This is where the challenge of the soft factors in business comes into play in implementing business optimization (BO). Humans essentially provide the cognition in AI systems. The volume of data and the speed of computing are such that once the algorithms are coded, machines can execute and humans cannot keep up. This can lead to a loss of control over AI systems. AI-based systems should not make automated decisions in situations where overseeing that decision by a human is important. Therefore, in this discussion on BO, NI is given substantial importance. Humanization is the complete balancing act of AI-based decision-making with NI.

Subjective customer thinking

Faster and more accurate decision-making leads to greater customer satisfaction and therefore greater customer value. Optimizing the value based on data analytics and predictions resulting from AI is, however, a subjective process. Customers and other users (e.g., staff) are humans whose needs and underlying context can change depending on myriad subjective factors. AI cannot ascertain all those factors beforehand and, therefore, is unable to sufficiently code the emotions, perceptions, impressions, and judgments made by humans.

Customers may make their decisions based on options and, perhaps, not always on rationality. AI algorithms are not designed to ascertain these subjective decisions and, therefore, need humanization. Humanization is the introduction of subjective elements in AI-based decision-making. This subjectivity is not limited to the business decisions, but also purchase and recommendation decisions by the customer. The subjective factors in providing customer value are:

- Is the decision providing something more worthwhile to the customer than the ability of the system to measure it?
- Is the decision ethically and morally appropriate to the given situation?
- Is the source of the data understood by the business? Are there possibilities of unethical sourcing of data and its use in analytics?
- What is the possibility of data bias? Is the input data skewed because of previous results but the current context has changed?
- Is it possible to weed out data bias by the use of ML algorithms?
- Is the decision in accordance with the law of the land? And have the legal considerations been included in the AI code?

- Is the decision right in terms of time and location? Timing corresponding to a situation can also be subjective (e.g., slow service in a fine dining restaurant is preferred for an anniversary dinner versus service in a fast food one).
- Is the data and decision made with full respect to the security and privacy of the customer?
- Is there a balance maintained between the corporate profit goals and the customer's well-being? Pursuing profit goals alone, no matter how legal, can backfire in terms of customer sentiments.
- Is the AI a black box with no opportunity of seeing what is inside? How is the situation of providing possible explanations addressed?

Most of the above evaluators of decisions are not quantifiable. These are subjective constraints leading to subjective customer value. Optimizing these decisions requires agility in decision-making. Agile characteristics in AI enable iterative and incremental decisions. These iterations facilitate the incorporation of consequence in the subsequent decisions. ML has its inherent limitations, and it provides analytical results based on extensive correlations. Furthermore, the depth and complexity of AI-based systems result in them becoming a black box. Human experience, intuition, knowledge, and expertise judge if the decision to be made is right, is in the interest of the society, and is ethically and legally sound. These decisions are the input for the next iteration of decisions.

AI compliments NI

AI is a tool for business. AI computes vast amounts of data using machine power leading to an understanding of patterns in data. AI does not reflect empathy and understanding of humans. AI can only analyze that which can be encoded. AI uses machines with learning capabilities which are themselves coded. The architecture of AI solutions has multiple layers of patterns (DL) that attempt to replicate human thinking once a pathway into that thinking is established. AI augments natural (human) intelligence, but does not replace it. AI is a software tool whose limitations can be mitigated and complemented by NI.

NI relates to adaptive learning by experience. NI has multiple layers in its depths that can handle complex and delicate decision-making. DL algorithms lack common sense. While AI can recognize patterns in vast datasets, there is no understanding of the meaning behind the pattern. A trend in weather (based on temperatures) and a corresponding trend in the temperature of a factory furnace are both datasets for AI. The analytics executed to make predictions are not interpreted by the system in their context. Providing checks and balances in AI is crucial and one way of doing that is to let ML algorithms identify multiple "what-if" scenarios that can be made to reason with each other.

AI is useful when it leverages uniquely human skills rather than attempting to replace them. Human skills and values are brought to the fore when judiciously combined with AI. For example, AI does not replace creativity and leadership which are still essential for business. Humans make decisions based on context. AI-based business processes need programming based on context. But the context keeps changing based on customer sentiments and needs. Therefore, NI and AI are both needed for a healthy business decision-making scenario.

KNOWN–UNKNOWN MATRIX FOR AI vs NI

DL mimics the human brain with its multiple levels and depths. As a result, DL algorithms are able to recognize speech (e.g., "Hey Google! Hey Siri!") and images (e.g., face recognition to open a cellphone). NLP and DL handle situations that are defined from the "known" aspect of business reality.

Creative thinking and problem-solving are essential human traits that can be augmented by AI but cannot be replaced. It is the "unknown–unknown" quadrant that is the most challenging to handle.

The following is an explanation of the unknown–unknown matrix shown in Figure 10.1.

Automation: Hard, mono-dimensional data

Successful automation happens only with simple and linear processes working on mono-dimensional data. Data is produced by human activities. The IoT sensors capture data based on configurations by humans. AI algorithms identify patterns in these datasets based on the instructions. Machine

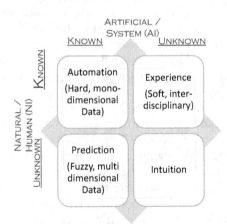

Figure 10.1 Known–unknown matrix for AI versus NI.

automated processes are isolated from the surroundings. ML algorithms do not need to understand the surroundings, nor the context in which they are operating. The automated processes faithfully operate as small components or packages, without any knowledge or understanding of how their functionality contributes to or supports the entire system or assembly. The knowns in the human arena are taken over by AI to automate repeated and monotonous tasks. Intelligence in machines reduces the onus of conducting repetitive tasks. Chatbots, robots, and digital trains are AI-driven technologies that can undertake many routine tasks. Automation has a higher speed of execution and accuracy. ML algorithms learn by storing the results of their decisions and, when presented with the same input parameters, arriving at the decisions much faster than humans. Nevertheless, these are human-like decisions and not human decisions.

Manual and routine tasks that are well defined are subject to automation. Automation needs humanization during execution. Optimization also needs humanization which has to be incorporated during design. For example, chatbots at various levels can answer queries on flights, play music, and provide initial health diagnoses. Learning by machines results in time and accuracy advantages. The design of the optimized processes incorporates these AI advantages but keeps provisions for human inputs. Besides, the consequences of decisions are evaluated by humans and fed back to the decision engine to enhance its data and code to be able to handle a similar context.

AI helps in identifying sale patterns and prioritizing resources needed to bring about a transaction. Prioritization of effort based on the likelihood of a particular outcome reduces human effort dramatically. The decision-maker needs vigilance in using the analytics because the prioritization provided by AI is based on past data. If the data is skewed for any reason and the AI logic is not able to figure it out, then the entire identification of pattern and ensuring recommendations could be analytically correct but realistically wrong. These kinds of situations need NI.

AI can be confused by new experiences. Emotions that cannot be coded or that change slightly from the ones that can be coded can throw an AI-based system off balance. Furthermore, unstructured data (which is a characteristic of Big Data) needs to be brought in some structured form before it can be analyzed. Coding the human experience relieves humans from putting effort to solve the same problems again provided it is the same experience occurring again. While ML is meant to "learn" from the experience, it is still the algorithm that provides the instructions to learn. Deciphering completely new experiences is outside the scope of AI.

Experience: Soft, inter-disciplinary

AI helps in identifying sale patterns and prioritizing resources needed to bring about a transaction. Prioritization of effort based on the likelihood of a particular outcome reduces human effort dramatically. The decision-maker needs vigilance in using the analytics because the prioritization provided by AI is based on past data. If the data is skewed for any reason and the AI logic is not able to figure it out, then the entire identification of pattern and ensuring recommendations could be analytically correct but realistically wrong. These kinds of situations need human experience.

Humans accumulate experience over time when they deal with a task. This experience helps relate several disparate tasks, understand their context, and generate creative thinking and solving problems.

Prediction: Fuzzy, multidimensional data

Machines are speedier in crunching large quantities of data enabling them to spot trends and make predictions. Machines take in multidimensional data, run algorithms through a large number of cycles, and dig out patterns in the data which are impossible for humans to identify. The extracting information and knowledge from vast multidimensional data by DL algorithms is beyond natural intelligence. ML (especially unsupervised) can help businesses ask the right questions. When the context is stable, ML can identify KPIs to help focus human decisions. But with changing context, NI is invaluable in arriving at the right decisions.

Intuition

Intuition is an outstanding feature of humans. Intuition, which comes from knowledge, long years of practice, and experience, is a crucial ingredient for NI. Intuition leading to solving problems is subtle. It cannot be precisely defined and is unknown to humans themselves. People can come up with completely new ideas and, at times, arrive at conclusions much faster than machines. ML by its very definition cannot reason abstractly and generalize. Physicians, artists, and musicians, for example, perform their art intuitively. Business decision-making has to make provision for intuition. Thus, the decisions can be initially made by NI and then scaled up accurately by AI. Alternatively, AI suggests a decision that is ratified by NI before it is scaled up. Constant cross-checking of the context is also mandated by respecting intuition in decision-making. AI advantage is limited if it is not combined with NI. People add valuable insights to decisions.

ADDITIONAL CHALLENGES IN DECISION-MAKING

These additional challenges form the basis for the need to superimpose NI over AI in decision-making. These challenges start with the DL architecture, which is a part of AI. This is followed by ethical, legal, and user experience challenges.

Deep learning (DL) challenges

DL, as discussed in earlier chapters, classifies data and identifies identify trends and patterns within that data. DL architecture (inputs, outputs, nodes, and layers) is a neural network that reflects the human brain and its multilayered decision-making capabilities. Similar to the brain, the DL backpropagation[4] algorithm assigns different weights to nodes in analyzing speech, images, and translations. As a result, DL provides insights beyond human capacity to enhance customer experience, speech and face recognition, driving autonomous vehicles, computer vision, and so on. DL's advances are the product of pattern recognition: neural networks memorize classes of things and more-or-less reliably know when they encounter them again. However, classification is not the same as human intuition, cognition, and contextualization.

This lack of DL to contextualize presents interesting challenges. "Teaching machines to use data to learn and behave intelligently raises a number of difficult issues for society."[5] Since AI is more than automating the existing tasks, there is storage of "learnings" from an experience of an interaction with a customer or solving a business problem. Machines can continue to incrementally learn to a level where the logic behind the learning becomes so deep as to be unexplainable. This is the situation where DL needs NI input.

DL is considered as resource hungry, unexplainable, and breaks easily.[6] This is so because DL needs huge training datasets that consume phenomenal resources, unexplainable due to deep multilayers, and breaks because it does not fully understand the context in which the decisions are being made. For example, "A robot can learn to pick up a bottle, but if it has to pick up a cup, it starts from scratch."[7] It not the lack of training and provisioning of depth that is the issue but the fact that DL (as a part of AI) cannot contextualize the situation which is continuously changing.

Ethical challenges of AI-based decisions

The aforementioned challenge of lack of contextualization leads to situations where the ethics and morality of decisions come into play. Straightforward predictions based on a clean set of data are most helpful in decision-making. These are the type of decisions that are automated. However, the uncertainty of the context in which the decisions are made presents a risk to automating AI-based decisions. Fully automated decisions which leave humans completely out of the loop and which are meant to provide customer value are risky. Customers have a subjective interpretation of their needs and the changing nature of their values. Full automation especially in customer-facing decisions may even be detrimental to business goals and to society in general.

The ethical challenges in AI-based decisions arise because of potential biases. AI systems make decisions based on data the provided and algorithms coded – both are subject to biases. While data is usually considered objective, it can still be biased since it incorporates the beliefs, purposes, biases, and pragmatics of those who designed the data collection systems. Data is not a singular record but a collection of many records of observations. Therefore, the potential exists that the beliefs of the observers have colored the meaning of the data.[8] Sample bias, prejudicial bias, exclusion bias, measurement bias, noise bias, and accidental bias are examples of data-specific biases[9] that influence the models built upon the data. Decisions to buy, sell, promote, and cut production are all sensitive to biases. Biases in the data and the opacity of the algorithms used to learn from the biased data are the central issues in AI and Big Data ethics.[10]

AI models can potentially code prejudices and beliefs. To find those biases requires careful auditing of the models, which only an NI superimposed on AI can handle (Figure 10.2). Appropriate checks and balances need to be put in place to prevent misuse of decision-making systems that rely on ML.[11]

Legal issues in unexplained AI

A human understandable explanation for an AI decision is also imperative from a legal perspective. Explainable AI provides a reason or justification for the analytics generated. The need to demonstrate the reasons for the analytical insights arises from the need to prove that the insights are not violating the legal frameworks of the region. The analytics are based on the relationship of data within the AI-based system. These systems are designed and owned by the developers. The algorithms are coded to enable them to traverse large dataset and establish correlations. There is no onus on the

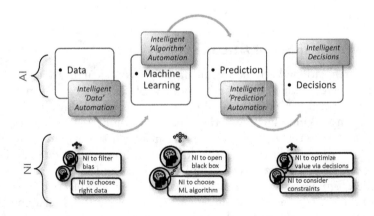

Figure 10.2 NI superimposing on the AI learning process in order to improve decision-making.

system to explain its decisions. An understanding of the data features and the high-level system architecture may still not be enough to explain or justify a particular recommendation.

These legal situations can have serious repercussions on BO. Disparate impact resulting from the decisions can lead to legal wrangling and court suits. Agrawal, et al. recommend examining the results from the analytics. "Do men get different results than women? Do Hispanics get different results than others? What about the elderly or the disabled? Do these different results limit their opportunities?"[12]

Incorporating NI in decisions is a way to ameliorate the impact of legally poor decisions.

Interfacing with humans

An important "soft" issue with AI-based systems is the way in which they interface with humans. Human-Computer Interface (HCI) is a discipline in its own right, encompassing interface designs, presentations, communication channels, and the growing expertise of the user. A static user interface will now "grow" with the expertise of the user and, therefore, may hinder her use of the system. Sight, sound, and touch are the basics of user interface design.

Interfacing with humans is an important element of successful BO.[13] As businesses relate to the users through multiple channels, the design of a website or a mobile app that provides analytics to the users needs to incorporate user experience.

Understanding the purpose of the customer's interaction with the business, modeling the processes, and reviewing multiple aspects of a user's (and user group's) relationship with a business help in improving the value of AI to the customer. User experience design is a specialist business analysis (BA) activity that makes provision for the incorporation of NI at all levels of the customer's interaction with the business.

SUPERIMPOSING NI ON AI

Nexus between NI and AI yields balanced intelligence in optimization. A judicious superimposition of NI on every stage of the AI ML pipeline is imperative for value-based decisions. Figure 10.2 shows the AI pipeline, with four phases: data collection, ML, prediction, and decision-making. The first three phases are relatively easy to automate based on current AI technologies as they can be defined. The fourth phase is not easy to be automated. The limitations of AI are handled by superimposing NI through the design, development, and implementation of the solution. The following outlines the role of NI in each phase:

- Data collection: choosing the right kind of data for a given ML problem and filtering the varied types of possible biases from the data
- ML: allocating the right kind of ML algorithm
- Prediction: opening the ML black box to explain causal relationships among inputs and prediction
- Decision-making: fully engaging in decision-making

Quality decisions, which are also ethical decisions, include humans in the decision-making loop. Humans are capable of considering the consequences of decisions vis-à-vis their quality and ethical ramifications. NI provides invaluable insights, after inspecting the consequences of decisions, by considering ethics and values. These NI-based insights are superimposed on the learning algorithm (as shown in Figure 10.2). The feedback loop illustrated in Figure 10.2 then tweaks the historical data, learning model, and new data to filter possible sources of error and bias and retrains the model. The learning-correction-relearning cycle is repeated multiple times to enable the system to continue to learn and improve its performance. Eventually, after multiple iterations, the model shown in Figure 10.2 arrives at ethically sound decisions that produce adequate customer value. The caveat to keep in mind is that in earlier iterations of this model, NI makes the actual decision, whereas in later iterations, AI learns from NI and stores those insights.

At an organizational level, as business processes are reengineered, a suite of principles related to ethics and morals can be adopted by the developers of the solutions. Visibility of the solutions through a walkthrough of models explaining the decision-making process, safeguarding the ingestion of data and its usage, and enabling judicious mixing of NI (humanization) in the decision-making process can go a long way in building trust in AI-based decision-making.

In Table 10.1, 1 addresses the bias issue, 2 and 3 address the inexplainability of AI models, and 4–7 address the inability of AI models for making decisions that have subjective value.

Biological neural network models[14] are discussed to help understand autonomous adaptive intelligence.

AGILE ITERATIONS ENHANCE VALUES

Critical thinking and problem-solving with AI

Critical thinking and problem-solving are human traits that are supported by rich data and analytics. AI ranges from general-purpose analytics (e.g., on historical, descriptive) to specific, fine-granular analytics (predictive). In each case, AI needs as well as supports critical thinking and problem-solving in business. AI can be used to simulate business scenarios. "Digital twins"

Table 10.1 AI limitations and NI superimposition over AI limitations for intelligent automation

AI limitation	Description	NI superimposition
Biases in AI models	AI models can only be as good as the data fed to them and the algorithms coded. Biases in models can crop-up based on the basis of observations and data, and those based on the developer's viewpoints.	NI challenges the data and algorithm biases mainly because NI is *not* limited to data and algorithms.
Inexplainability of AI models	AI models are a "black box" in which a large amount of data is fed and results come out. Feedback in AI models is also made objective.	NI helps in understanding the underlying causes of decisions.
Complexity	AI models are extremely complex – that are difficult to troubleshoot.	NI brings in intuition, experience, expertise, and associated knowledge.
Performance-driven metrics	AI models base their successes on performance-driven metrics. This leads to ongoing optimization that may not care for value.	NI seizes the opportunity to vary the performance-driven metrics based on the needs of the time.
Ethics and moral not codable	AI models can only encode the well-defined processes, and they can only analyze data that is available.	NI can superimpose ethical and moral values based on the context of the situation.
Values are context-driven	AI models can understand the context only to an extent that the context can be coded. If the change in context is not describable or visible to the AI models, that context is lost to the model.	NI is in a position to understand the context much better than AI – because NI is capable of absorbing contradictions and misalignments in values.
Sequential vs agile	AI models are sequential – moving from manual to automated to optimized processes.	NI, superimposed on AI, can make processes increasingly Agile.

used in simulating dams, human bodies, and hurricanes can also be used to simulate the trends and pathways of the business.

Critical thinking is undertaken in an iterative and incremental manner within BO. Critical thinking approaches a problem in a disciplined manner. Critical thinking starts by conceptualizing a problem, followed by analyzing it. AI-based analytics are immensely helpful in the analysis of the problem as they provide insights that complement NI. NI supports critical thinking by enabling an understanding of the changing subject matter or the context in which a problem is occurring. Both the problem and the solution are subject to this changing context which AI may not be able to decipher.

An important development to support critical thinking is the Hex-E protocol,[15] which explores machine learning models in an iterative and incremental way. Superimposition of NI on Hex-E is facilitated by AI. For example, Hex-E facilitates a backpropagation algorithm using through its automated protocols that can also be explained. Hex-E protocol, with automated correlations, can enable and support creativity, help solve problems by recombining ideas, and develop fundamental new interface primitives.

Decision– action–decision–feedback cycle

Agility in business decision-making is based on iterations and increments. NI plays an important role in these iterative and incremental decisions because it provides input in the consequences of the decisions.

Table 10.2 shows this Agility and its incremental improvement for auto-mation and optimization. These iterations insert AI with due respect to AI and as a means to help human decision-making abilities and incorporate the impact of consequences in a virtuous feedback cycle.

Table 10.2 The decision–action–decision–feedback cycle for optimization of business processes with inputs from NI

	Design and develop	Execute and make decisions	Iterate after examining consequences
Automate	Create a model that replicates exactly what humans do with no variation and code that.	Let machines execute the algorithms with varying data. Machines can only vary the execution based on the parameters and the data input.	Slightest variation in input can potentially change the way the machine understands it and throws the results out of balance/in chaos.
Optimize	Reengineer business processes by questioning each activity for its contribution to the overall goal of the process. Model and code with flexibility and encapsulation in mind.	Machines execute code only to the extent there is no unexpected variation in the input. Humans are overseeing and providing relevant inputs to the business process execution.	Agile characteristics of iterations and increments are incorporated in the iterations; decisions and their consequences are evaluated based on human-values (ethics, morality, and legality) and fed back in the system.

CONSOLIDATION WORKSHOP

1. Discuss a situation you are familiar with in the context of the known–unknown matrix? Include examples of where AI plays an important role through automation and where NI is even more important than AI?
2. Why is pure automation a bad idea in business systems? Discuss the limitations of AI from human perspectives? *(Hint: include ethics, morality, and legality of decisions)*
3. Explain the various kinds of biases that can creep in data. How can such biases be filtered?
4. What is algorithmic bias? How can it be detected and minimized?
5. Discuss with examples the ethical and moral challenges of AI-based decision-making.
6. Why is the explainability of AI systems important? What happens with the "black box" AI?
7. Describe the role of subjective human values and its importance in creating customer value.
8. What is critical thinking and problem-solving capability? How can an organization strike the right balance between AI and NI through critical thinking?
9. Are machines capable of critical thinking? Discuss both sides of the question with arguments and examples.
10. AI systems can support and bring the decision-maker to the decision point beyond which a judicious combination of NI with AI is necessary. Why? How can it be brought about by the decision-action-decision-feedback cycle?

NOTES

1. Agrawal, A., Gans, J., and Goldfarb, A. *Prediction Machines: The Simple Economics of Artificial Intelligence*, Harvard Business Review Press, Boston, MA, 2018.
2. Lovelace, A. Augusta Ada King, Countess of Lovelace was an English math-ematician and writer, chiefly known for her work on Charles Babbage's pro-posed mechanical general-purpose computer, the Analytical Engine; https://en.wikipedia.org/wiki/Ada_Lovelace accessed 4 Nov, 2020.
3. See note 1.
4. Hinton, G. A professor emeritus at the University of Toronto and the grand-father of backpropagation, who sees "no evidence" of a looming obstacle.
5. Hull, J. *Machine Learning in Business*, University of Toronto, © J. Hull, Chapter 8, pp. 161–170 "Issues for Society", Toronto, CA, 2019.
6. Based on Jason Pontin, Greedy, Brittle, Opaque, and Shallow---The Downsides to Deep Learning *IDEAS*, 02.02.2018. https://www.wired.com/story/greedy--brittle-opaque-and-shallow-the-downsides-to-deep-learning/.

7. Unhelkar, V. V., Li, S., and Shah, J. A. Semi-supervised learning of decision-making models for human-robot collaboration. In the *Conference on Robot Learning*, Osaka, (CoRL 2019).

8. Provost, F., and Fawcett, T. *Data Science for Business: What You Need to Know about Data Mining and Data-Analytic Thinking.* O'Reilly Media, Newton, MA 2013.

9. Tim, J. M. Machine learning and bias---look at the impact of bias and explore ways of eliminating bias from machine learning models. *IBM Developer*, 2019 Aug, 27.

10. Ethics of Artificial Intelligence and Robotics, Stanford Encyclopedia of Philosophy, 2020, April 30. https://plato.stanford.edu/entries/ethics-ai/#OpacAISyst.

11. Finlay, s., Ch 13, p. 111.

12. Agrawal, A. et al. Prediction machines, Ch 18, p. 197.

13. Unhelkar, B. User *Experience Analysis Framework: From Usability to Social Media Networks* – Cutter Executive Report (12,000+ words), Data Insights and Social BI, Vol. 13, No. 3, Boston, MA, 2013 Apr.

14. A path toward explainable AI and autonomous adaptive intelligence---deep learning, adaptive resonance, and models of perception, emotion, and action Stephen Grossberg, *Frontiers in Neurorobotics*, Vol. 14 No. 2020, Article No. 36, 26 p. 2020 Jun.

15. Unhelkar, B., and Nair, G. Embedding Intelligence within data points for a Machine Learning Framework---Hex-Elementization. *Presentation and Proceedings of the IntelliSys 2019 Conference*, London, 2019, 5–6 Sep.

Chapter 11

Investing in the future technology of self-driving vehicles

Case study

INTRODUCTION

About 1.35 million people die in road crashes each year and an additional 20–50 million suffer nonfatal injuries, often resulting in long-term disabilities. On an average, every 24 minutes somebody is knocked down dead by a human driver.[1] Safety in driving is one of the most important contributions of artificial intelligence (AI) in the modern world. Many lives are saved through the AI technologies everyday on the roads. Unfortunately, a handful of fatal crashes involving autonomous vehicles (AVs) have received publicity that dissuades people from contemplating a self-driving car. Official approval by the regulatory agencies after rigorous safety testing is not able to overcome the human element in the application of AI to autonomous driving.

Public awareness on self-driving vehicles is largely from hearsay or newscast. Facts are different. The truth is, a true unmanned fully developed self-driving car is far away. What is heard in newscasts and read in newspapers is about vehicles which are in various stages of self-driving development. They are in an experimental stage and occasional accidents are bound to happen. Some studies suggest that self-driving vehicles will probably have to go through not just millions, but billions of miles of testing to win the trust of the public.

For the first time in 130 years, the world is in the midst of a major transformation in automobile transportation.[2] The advent of driverless cars is going to revolutionize the way people get around. While the technology is likely to bring benefits to society, there will also be unintended consequences to consider. The potential loss of driving jobs along with other jobs related to driving such as insurance and a severe blow to the traditional auto industry are some of the more obvious negative consequences. The technology is already picking up steam and is on the verge of making a paradigm shift in the way people think about cars and mobility. It offers opportunity to the public to consider the benefits and get rid of unfounded fears, challenge to governments and policymakers to frame new traffic laws, and to entrepreneurs and investors a new avenue for investment.

This is a case study on self-driving or Autonomous Vehicles (AV) which are making news headlines these days. Of paramount importance to the understanding of the levels of self-driving cars is the conceptual model proposed by the Society of Automotive Engineers (SAE) International describing the various levels of autonomous driving. The case study describes these levels in detail, explains the engineering and technology behind AVs, states their benefits over human driving, and estimates the impact on economy if the emerging AV technology is pressed into service.

PUBLIC AWARENESS OF AUTONOMOUS DRIVING TECHNOLOGY

What is the level of public awareness regarding self-driving cars? How safe does the public think self-driving or AVs are compared to their counterpart human drivers? Is the emerging technology here to stay? Table 11.1 shows a public survey conducted in the USA on AV technology in early 2020.

The survey concludes that, "consumer acceptance of automated vehicles seems to be stuck in neutral."[3] In other words, most of the people who responded to the survey are neither strongly for nor strongly against the automated driving technology. Almost half of them would like to get some more information on the technology and the legal aspects of AV driving. This applies to the entrepreneurial circles, too.

Most automakers are carrying on development and testing in the secrecy of their R&D labs or on testbed driving circuits hidden from public view. When they have to put to test their latest AV models on the roads, they are vividly decorated or heavily camouflaged, because "it is important to keep future products secret, to avoid undercutting the sales of current products

Table 11.1 Public survey on AV technology

AV aspect	%
Riding in a self-driving car	
• Trust	12
• Not sure	28
Safety	
• Safer if one has control in emergency	72
• Safer if there is a human backup	69
• Safer knowing AVs have passed rigorous testing	47
• Feel safer after seeing a demo	42
Like to see more information on AVs	
• Legal responsibility of crashes	57
• Laws regulating AV safety	51
• Vulnerability to hacking	49
• Easy to understand explanation of how AVs function	44

and to build anticipation of the new product."[4] While there is an exponentially rising number of journal papers and conferences on AI, ML, DL, and their applications, very few are on their applications to AVs. The few precious ones are mostly from academia who develop AVs driving simulation models and software platforms in their labs.[5,6,7]

Not much is known about the engineering and functionality of AVs. It is human psychology to fear the unknown. This fear is further fueled by media reports on the crashes involving AVs, which often hit the newspaper headlines. The fact is AV technology is far from being mature. There are several developmental and operational AV levels as described in the following section. The reported crashes have occurred, fortunately, only at a development and/or testing level of the fast-evolving AV technology.

SAE LEVELS OF AUTONOMOUS DRIVING

The Society of Automotive Engineers (SAE) International has proposed a conceptual model defining 6 levels of driving automation which ranges from fully manual level 0 to fully autonomous level 5. The model has become the de facto global standard adopted by stakeholders in the automated vehicle industry (Figure 11.1).[8]

Level 0: No automation

Most driving today is Level 0 automation driving in which the human driver maneuvers the vehicle in all road conditions. Even vehicles equipped with an automatic transmission system (as opposed to manually changing gears) and emergency braking system are classified as belonging to level 0 automation because such automated systems are passive and do not drive the vehicle. The dynamic driving task is fully performed by the human driver.

Level 1: Driver assistance

This is the lowest level of automation. The vehicle features a single automated system for driver assistance, such as cruise control. The cruise control

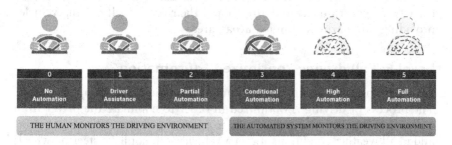

Figure 11.1 SAE levels of autonomous driving.

system aids the human driver in maintaining a steady speed and keeping it at a safe distance behind the traffic in front. The cruise control system also has the subfunctions of de-acceleration and braking when it closes the gap between itself and the vehicle in front and of re-acceleration when the gap widens to more than the designated safe distance. The human driver is responsible for controlling the rest of the driving functions.

Level 2: Partial driving automation

The vehicle is equipped with Advanced Driver Assistance Systems (ADAS), which can control both steering and acceleration/deceleration and braking. However, there is always a human belted to the driver's seat, ready to take control at any time. Tesla Autopilot, General Motors Cadillac Super Cruise system, Volvo Pilot Assist, and Nissan ProPILOT Assist are some of the recent level 2 driving systems.

Level 3: Conditional driving automation

Level 3 automation level is a sudden jump from level 2. Vehicles are equipped with driving environment detection capabilities. They can make informed decisions for themselves, such as accelerating past a slow-moving vehicle. However, the driver must remain alert and ready to take control if the system is unable to execute the task. Moreover, the human driver can always override the decisions made by the level 3 vehicle. Honda Legend Sedan and Audi A8 are the latest level 3 cars hitting the market.

Level 4: High driving automation

Level 4 vehicles can intervene if things go wrong or there is a system failure. In this sense, these cars do not require human interaction in most circumstances. However, a human still has the option to manually override. Level 4 vehicles can operate in self-driving mode. But until legislation and infrastructure are developed, they can only do so within a limited area (usually an urban environment where top speeds reach an average of 30 mph). This is known as geofencing. Some automakers have signed contracts with governments and already put their level 4 self-driving vehicles providing ridesharing in limited urban areas.

Level 5: Full driving "optimized" automation

Level 5 automation eliminates the human from the driving loop altogether. The process of driving is optimized in order to eliminate the driver from actual driving. The vehicles are able to drive anywhere a human can drive and in all weather conditions. Level 5 vehicles will not have steering wheels or acceleration/braking pedals. Fully autonomous cars are undergoing testing in several pockets of the world, but none are yet available to the general public.

Benefits of autonomous driving

The economic and social benefits of autonomous driving are so vast that even the skeptic who sees and understands these benefits, will agree.[9] This section deals with some of the major benefits to humanity that AVs will deliver if they are adopted on a large scale.

Safety

Currently, the popular belief seems to be that human drivers are much safer than AI-driven road vehicles. However, statistics seem to indicate otherwise. The Association for Safe International Road Travel (ASIRT), a nonprofit, nongovernmental organization that seeks to improve the personal safety of travelers on the roads, has published the annual global road crash statistics on its website (Box 11.1).[10]

What are some of the causes of human-driven accidents and fatal crashes?

A 2015 study by the US Department of Transportation NHTSA attributed 94% of accidents to human error, with only 2% to vehicular malfunction, 2% to environmental factors, and 2% to unknown causes.[11] This increase in safety also further reduces the cost of secondary factors such as emergency and medical services, government roadway management, and beyond.

Drunken driving, distracted driving often due to smartphone addiction, and old age driving are additional causes of accidents. According to the statistics released by the Japanese traffic authorities, in 2019, the number

BOX 11.1   DATA OF ROAD ACCIDENTS CAUSED BY HUMAN DRIVERS

Approximately 1.35 million people die in road crashes each year.

On average, 3,700 people lose their lives every day on the roads.

An additional 20–50 million suffer nonfatal injuries, often resulting in long-term disabilities.

More than half of all road traffic deaths occur among vulnerable road users – pedestrians, cyclists, and motorcyclists.

Road traffic injuries are the leading cause of death among young people aged 5–29.

On average, road crashes cost countries 3% of their gross domestic product.

of fatal accidents caused by drivers aged 75 or older accounted for about 15% of the total. With the rapid aging of Japan's population, the number of people 75 or older who hold driver's licenses increased to 5.63 million at the end of 2019. It is estimated the figure will reach 6.6 million in 2022. The Tokyo Metropolitan Government is reportedly considering subsidizing the cost of attaching safety and drive assisting devices to the cars of elderly drivers.[12]

The US National Highway Traffic Safety Administration (NHSTA) declares that the Advanced Driver Assistance Systems (ADAS) available in present-day vehicles that help drivers from drifting into adjacent lanes or making unsafe lane changes, or braking automatically if a vehicle ahead of them stops or slows suddenly, among other things are already helping to save lives and prevent injuries.

As stated above, autonomous driving is still far from being mature. It is expected that the future AV technology will be fully equipped with many redundant safety features which will drastically reduce road accidents and fatalities. In summary, some of the benefits will be:

Congestion

Congestion in the cities and on the highways is a daily phenomenon which has become a two-way nightmare for officegoers. Vehicles caught in congestion lead to a great deal of lost time and fuel. According to the INRIX 2019 global traffic scorecard which is the most in-depth congestion and mobility study of its kind, Bogota in Columbia is the most congested city in the world where drivers lose 191 hours (almost 8 days) per year. Most US cities are also highly congested costing each Americans nearly 100 hours equivalent to $1,400 a year.[13]

Self-driving cars will drive efficiently and through communication with one another will maintain a constant flow of traffic. Fewer traffic jams will save fuel and reduce greenhouse gases from needless idling.[14] As an added benefit, reduction in congestion will also reduce human annoyance, boredom, and stress.

Pollution

Current automobile technology spends far more fuel than necessary to reach from point A to point B. The environmental pollution resulting from fossil fuel combustion is huge. Current self-driving car models are basically gas-driven cars with gas and brake pedals and steering wheels with Advanced Driver Assisting Systems (ADAS) fittings. However, as autonomous driving reaches level 5, the concept of the car itself will change. Fully automated vehicles will be electrically driven[15] and fitted with electronic control systems. Gas and brake pedals and steering wheels will become obsolete. Without any exhaust tailpipes and emissions, fully automated electric vehicles will greatly reduce environmental pollution.

Parking space

Every personal car that transports a person to work, shopping, entertainment, or whatever needs parking space. Cities and urban areas are flooded with parking lots which are always almost packed with cars so much so that drivers must be constantly on the lookout for vacant parking spots. Self-driving cars will be programmed to let the passengers off at the front door of their destination and quickly move away to serve other passengers. Keeping the self-driving cars continually in motion will lead to a reduction in the parking space. The land freed from parking lots can be utilized for other city development purposes.

Passenger quality of life

All the occupants of the self-driving vehicle will be passengers. While traveling, they could engage in productive work like reading, attending to emails, or entertainment activities like watching movies. The elderly and people with disabilities find it very difficult to move from place to place using the current private and public transportation system. They will be able to enhance their mobility given the availability of self-driving cars. They will call a self-driving vehicle from the nearby fleet and get a ride to wherever they want to quite comfortably. Parents will not have to worry about getting their kids to school in the morning and picking them up after school. Secure self-driving cars can be programmed to transport the kids to and back from school.

Cost benefits

Self-driving vehicles have not yet reached a visible maturity level because of which it is difficult to project the cost benefits of this new mobility technology. Currently, owning individual AVs is quite expensive since the production of such cars has not yet hit the main mobility market. But several estimates of the cost benefits are underway.

Some studies estimate that self-driving cars will save the United States over $300 billion per year, mostly due to the reduction in road accidents. The cost benefits are still higher if the savings due to congestion relief, increased productivity while traveling, and the reduced cost of delivering goods and services are taken into consideration.[16]

The UK firm Thales estimates that AVs will bring 90% reduction in traffic deaths, 60% drop in harmful emissions, 100% elimination of stop-and-go waves that currently create traffic congestions, 10% improvement in fuel economy, 40% reduction in travel time, and a hard-to-believe 500% increase in lane capacity. The net savings on reduced insurance costs, reduced running costs, and parking mean the UK consumer market would save about 5 billion pound annually.[17]

According to expert projections, AVs will be safe and reliable by 2025 and may be commercially available in many areas by 2030. If they follow the pattern of previous vehicle technologies, during the 2030s and probably the 2040s, they will be expensive and limited in performance. They might occasionally require human intervention in emergency and unexpected situations. Figure 11.2 shows the comparative user cost estimate of AVs and HDs (human-driven vehicles). The cost is computed in dollars per mile transit. AVs are likely to cost more than HDs and vehicles and public transit, but less than HD taxis and ride-hailing services.[18]

Unintended consequences of automated cars technology

Automated driving will bring a host of benefits to humanity. The social and ethical issues surrounding self-driving cars are hitting the headlines, but in-depth research is yet to be done.[19,20] The unavoidable unintended consequences[21] of AV technology which is the source of social and ethical issues are discussed in this section.

Loss of jobs

Millions of people earn their living from driving taxis, hired cars, buses, and trucks. If AVs become a reality on the roads, these people will suddenly end up losing their jobs. According to the US Bureau of Labor Statistics, the

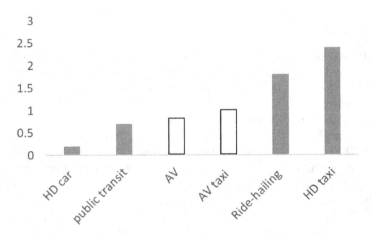

Figure 11.2 Comparative running cost of AVs and HDs ($/mile).

total number of potential jobs lost would be about 4.5 million. This is without taking into consideration the additional management and supporting staff employed in the driving sector. Moreover, the low-skilled driving force cannot be immediately retrained to take up new jobs. Retraining would further increase costs.

Blow to the auto industry

Not the traditional auto manufacturers, but IT companies like Google are at the forefront of the driverless-car technology. Hence, the success of driverless cars will directly hit the traditional automakers. Moreover, since the driverless cars will be easily available everywhere and readily transport people anywhere, less and less people will be inclined to own a car, given the headaches of finding a parking lot when driving and paying the taxes, repair, and maintenance bills. Thus, private car ownership will become a thing of the past, suffocating the automobile industry.

Blow to the auto insurance industry

Personal injury coverage, auto liability coverage, comprehensive coverage, collision coverage, medical payments coverage, and so on are the myriad insurance coverages routinely included in any car insurance package. Although not all of them are mandatory, most people end up purchasing more than the essential amount of coverages for fear of risk to life and property. Automobile insurance companies are thriving because of the numerous risk factors associated when humans are driving. Accident risks increase in proportion to the number of vehicles in use, which has been rising unabatedly over the years. Driverless cars promise to greatly reduce the occurrence of risks, which will impact the survival of the auto insurance companies. Lack of private ownership of cars and the elimination of potential risk in autonomous driving will knock down the traditional business model of auto insurance.

The bottom line is: The advent of driverless cars is going to disrupt and revolutionize the way people get around. While there is likely to be a net positive benefit to society, there will also be unintended consequences to consider. It is important to be prepared for these, and any other, unintended negative consequences that may materialize as a result of this disruptive technology.

It is a historical fact that the global economy did not fatally collapse under the disruptive tidal wave produced by the industrial revolution. Traditional jobs perished overnight, but new jobs were also created in the aftermath. Experts predict the rapid advances produced in AI will bring about a similar disruption in the global economy. But it will also create new jobs.[22,23]

AV ENGINEERING

This section aims at explaining the AV engineering to nontechnical readers. The best way to understand how AVs drive is by taking a closer look at how human beings drive. The subsection below presents an analysis of human driving; the subsection which follows explains AV driving which is an extension of human driving.

Analysis of the human driving cycle

A close analysis of the mental processing and physical activities involved in human driving reveals two kinds of cycles in operation – the foreground conscious cycle and the background unconscious cycle (Figure 11.3). Deeper analysis of these two cycles helps to better understand the pros and cons of autonomous driving.

Foreground conscious cycle

The foreground cycle of mental processes and physical activity takes place at the conscious level of the driver. The cycle goes through iteration of the following four major steps: Perception, scene generation, planning, and action – explained below in detail.

Perception

Perception refers to the taking in of all the information that impinges on the senses. The major sensing necessary for driving is visual. Through the

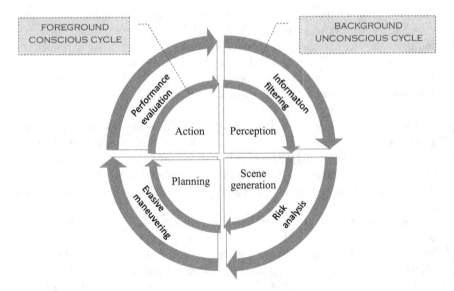

Figure 11.3 Human driving cycles.

eyes, humans perceive the traffic signals and their colors, read the traffic signs posted on the road, read the signs and words written on the road surface, and see the traffic moving in front. Small mirrors fitted to the car and large mirrors by the roadsides act as secondary means of aiding the driver's vision.

Scene generation

The information picked up by the sensory organs is transmitted to the brain via the nerves. The brain interprets the pieces of information supplied by each sense organ and combines them to build a world map of the surroundings in which the human driver's car happens to be at that point in time. From perception, humans generate the driving scene surrounding their vehicle. They pinpoint the other vehicles, cyclists, pedestrians, signals, road signs, and weather conditions and make an action plan.

Planning

From experience, the human driver how the situation will change the moment the signals controlling the flow of traffic through the crossing change. The still looking crossing will soon bustle with activity with traffic and pedestrians moving in all directions indicated by the green signals, while those patiently waiting as indicated by the red signals.

Action

Finally, all the planning is put into action, triggered by the signal change. The human driver will release the brakes, gently press the gas pedal, and steer towards the vehicle right. However, driving doesn't end after executing the planned action. The perceiving, scene generating, planning, and acting cycle iterates rapidly.

Background unconscious cycle

As the human driver is actively managing the foreground driving cycle, there is another subtle cycle operating at the background, which may not be at a fully conscious level. This cycle, too, can be broken down into four distinct steps.

Information filtering

When driving, a variety of external information not directly related to driving presents itself to the driver: for example, the scenery, buildings, shops, lampposts, and so on. A good driver concentrating on driving filters out the unnecessary information almost unconsciously.

Risk estimation

On scanning the driving environment, an experienced driver also estimates the risk factors before executing the plan of action. For example, when the car has stopped for the red signal at a crossing, the driver may notice a cyclist at the side or a child at the curb. If the driver has planned turning left with the left winkers already set blinking, she will think of the possibilities of the cyclist cutting straight as the car is turning or the child suddenly hopping on the pedestrian crossing.

Exception handling

With the risk estimation associated with the action planning, the driver is ready with an exception handling or emergency plan if the risks do materialize. In the above example, the driver will do a left turn extremely slowly while cautiously monitoring the moves of the cyclist and the child. She may also have to apply the brakes if, at the critical moment of turning, the move of the cyclist or the child is not in keeping with the appropriate traffic rules. This is exception handling which overrides the normal planned action.

Performance evaluation

After executing the plan of action, an experienced driver briefly evaluates her driving from perception to action. She may become aware that she was not paying attention to the pedestrians waiting to cross the road at the junction she has just passed. She had not stopped there to let the pedestrians cross the road. Performance evaluation heightens the awareness of the driver and leads to greater safety in driving.

In human driving, even if the perception followed by scenario generation, and action plan is perfect, on some occasions, there is an error in action. The human operator may mistake the gas pedal for the brake and end up accelerating the car unintentionally.

AV driving cycle

The very same steps are involved in an AV driving cycle.

Perception

In an AV, there are arrays of sensors fitted to the body of the automobile like cameras, radar, ultrasonic sensors, and LIDAR. The sensors constantly pick up the information of the surrounding world. The AV perception of the driving environment is through its sensors.

Ultrasonic sensors

The giant Japanese car manufacturer Toyota was the first to use ultrasonic sensors for parking assistance systems. The ultrasonic waves (which are above the audible frequency range of humans) transmitted by the gadget are collected on reflection from the surrounding objects. The transmitted and received signals help to calculate the distance of the object reflecting the waves from the sensor's position. Ultrasonic sensors are not adversely affected by poor weather conditions. They work well in detecting nearby objects around the vehicle, especially pedestrians. On the downside, ultrasonic sensors are sensitive to interference from high-frequency sounds emitted in the neighborhood.

Visual camera

Visual camera is the cheapest and the most common sensor used in almost all AVs. A host of visual cameras are attached to the vehicle chassis, each one performing a specific function. Two mono cameras for the front vision of the vehicle form a stereo vision system. They have relatively larger focal lengths to see as far as possible; the cameras on the side provide an immediate view of the objects located on both sides of the vehicle. The cameras on the rear end help in reversing and observing the traffic behind the vehicle. Since the sharpness in focus and clarity of present-day cameras have improved tremendously, these simple sensors are used to perform multi-function necessary for driving like recognition of signals, road signs, lanes, hazards, traffic, and pedestrians. Recognition is done by the state-of-the-art deep learning algorithms. However, cameras are very sensitive to light. Too much or too little of light will greatly hamper the visual information picking of the cameras.

Radar

Radar (*radio detection and ranging*) sensing using radio waves for object detection and tracking has been in use in navigation for a very long time. Radar emits radio wave pulses and using echo and Doppler effect calculates the size and distance of the objects reflecting the radio waves. It can also detect the direction and speed of moving objects. Because of its long-range applicability, radar is used primarily in adaptive cruise control and lane change assistance systems in AVs.

Lidar

Lidar (*light detection and ranging*) works on the same principle as radar. In place of radio waves, it uses laser light pulses for detecting objects surrounding the AV. The accuracy and high resolution of Lidar have made it an outstanding choice for 3D surveying and mapping in AVs. Although the current costs are exorbitant, Lidar is considered indispensable as an AV sensor. However, recent AV manufactures, notably Tesla, have demonstrated that there is alternative cheap sensor technology for 3D mapping.

Global positioning system

Global position system done via satellites orbiting the earth has become very common in driving. Its accuracy is approximately 3 m, which is sufficient for land driving purposes. The weak point is that it requires a clear line of sight from between the vehicle and the satellite on top. It does not function when the vehicle goes underground or in a tunnel.

Scene generation

This is where the latest state-of the-art computer vision deep learning algorithms come into play. Some computer vision deep learning algorithms are so advanced that their performance has exceeded that of human experts.[24] There is a fusion of all the information provided by the sensors to produce an instantaneous 3D environment around the AV.

Planning

All the decision-making and planning functions are controlled by the central computing platform. Real-time scene generation by fusion of sensor data requires massive data processing and storage. The real-time processing is done by graphic processing units (GPU) containing thousands of cores delivering massive parallel compute orders of magnitude higher than a multicore CPU. The Nvidia drive AGX PEGASUS[25] is one such platform. Designed for the future level 4 and level 5 autonomous driving and robotaxis, it can achieve an unprecedented 320 trillion operations per second (TOPS) of supercomputer.

Action

The action plan decided upon by the computing platform is transmitted to the car mechanical parts via the actuator interface. The perceive-plan-act driving cycle operates at the millisecond (one-thousandth of a second) pace.

Humans vs AVs driving

The planning, execution, and prediction steps involved in human driving closely match the steps that have been analyzed in the conscious and subconscious driving cycles that the authors have proposed in the preceding subsection. These steps will be coded as programs and embedded in the AV controlling software platform. They will be further refined through the process of deep learning carried on not by just one AV on the road, but literally by millions of them on the roads and billions in the simulation world. Since the machine perceive-plan-predict-execute activities occur in a couple of milliseconds compared to the fastest human judgment and activity that take a couple of seconds, machines are expected to have an upper hand in planning and executing fail-safe measures.

However, the following study disproves the claim that machines are unquestionably better than humans in driving. The Insurance Institute for Highway Safety (IIHS) recently examined more than 5,000 police-reported crashes across the USA and separated the driver-related factors that contributed to the crashes into the following five categories:[26]

- Sensing and perceiving errors: Poor visibility, failure to recognize hazards, and driver distraction
- Incapacitation: Impairment due to alcohol or drug use, medical problems, or falling asleep at the wheel
- Planning and deciding errors: Driving aggressively or too fast or too slow and not leaving enough following distance from the vehicle ahead
- Execution and performance errors: Inadequate evasive maneuvers, overcompensation, and other mistakes in controlling the vehicle
- Predicting errors: Incorrect estimate of the speed of other vehicles, and incorrect assumptions about what other road users are going to do

The study showed that crashes due to only sensing and perceiving errors accounted for 23% of the total and incapacitation accounted for 10%. The remaining two-thirds of the errors occur in planning, execution, and prediction.

Given the fact that AVs do not share human frailties like inebriation or distraction and assuming that the sensors guiding the future AVs are hundred percent foolproof, the IIHS study warns that AVs would only avoid about a third of the errors that lead to crashes. Thus, leaving humans out of the driving loop may never become a possibility. As argued in Chapter 10, NI superimposed on AI helps in making decisions in tight and extreme situations.

Of course, no machine can override the laws of physics as in the case when a vehicle in front of the AV suddenly stops or an animal or a pedestrian suddenly jumps in front of a speeding AV. Mechanical failures, too, cannot be completely avoided. Employing fleets of fully matured AVs for transportation may not reduce the number of accidents to zero, but the figures will not be any near to the annual 37,000 deaths and millions of injuries prevalent today.

THE STATE-OF-ART OF AVs ENGINEERING

Brief history of self-driving cars

Ever since the emergence of the automobile from the horse-driven carriage technological culture, there have been sporadic experiments and at times impressive demonstrations on public roads of autonomous driving vehicles. The first serious attempt at autonomous driving was by the US Defense Advanced Research Projects Agency (DARPA), the very same agency who had laid the foundation in communications that led to the invention of the internet. In 1985, DARPA developed the Autonomous Land Vehicle (ALV) that could self-drive on rough and dangerous terrain without an onboard human driver. A fleet of ALVs used closed-circuit camera to sense the surroundings and communicate with one another in transit. The sensing and processing technology of the time being rather primitive, the ALV fleet could travel no more than at snail's pace. Then came Carnegie Mellon's Navilab experiments which lead to the development of the well-known Autonomous Land Vehicle using a Neural Network (ALVINN) in 1989. This was the first AV to introduce an Artificial Neural Network software to process the car sensor's input. Because of the efficiency in the input sensor information processing, ALVINN could travel much faster than DARPA's ALV.

European car manufacturers were also researching on AVs at that time. In 1995, the VITA (Vision Information Technology Agency) project spearheaded by Mercedes-Benz had several demonstrations in Germany, France, and Denmark. In one of the recorded demonstrations, the AV covered a trip of more than 100 km traveling at speeds over 100 km/h. It was capable of maintaining itself within the lanes and also of changing the lanes to overtake slower cars on the highway.

In the 1990s, car manufactures began to equip their cars with onboard diagnostic systems that monitor the engine and other mechanical functions. The diagnostic and monitoring systems are essential in developing an integrated self-driving software platform.

In 2004, DARPA organized the first self-driving Grand Challenge. The competition offered a prize of one million dollars to any self-driving car that could finish a 142-mile complete course. The course was too rough and unpredictable for the AV technology of that time. However, in DARPA's Grand Challenge II in the following year, five teams finished the course. DARPA's Grand Challenge II became a milestone in the history of AVs which boosted the confidence of AV engineers and technocrats.[27]

The future of self-driving cars

AVs have come a long way from the Grand Challenge competitions. Embedding ADAS has become a commonplace in the latest car models, although at an additional expense. Lane-departure warning, lane change, signal detection, traffic sign recognition, road-marking recognition, hazard detection and avoidance, pedestrian detection, pedestrian movement detection, auto-cruise, collision detection, auto-braking, and self-parking assistance functions have greatly reduced stress and made driving comfortable to human drivers. However, a fully mature level 5 autonomous driving technology is still very far away.

Experts and visionaries call the self-driving cars project as a moonshot.[28] Imagine what an effort it was to launch a satellite in outer space, rocket man to the moon and back. Most people were skeptical about the technology of the time. But the effort, time, and money spent for the project turned out to be a big success. Self-driving technology is similar in stature to the lunar missions. It is about to make a paradigm shift in transportation far greater than the one from horse-drawn carriages to automobiles about a hundred years ago.

Progress in self-driving technology has to be incremental – one stage at a time. The SAE model is indeed the roadmap. Most people are skeptical about self-driving cars, because misinformed by the media hype they are made to believe that fully automated cars will become available within a couple of years. But the scientific and engineering truth is that the technology needed for a fully functional level 5 car is not yet ready. NI, not AI, is ruling the roads.

Two predominant difficulties should be overcome by the self-driving technology to become viable: technology maturity and cybersecurity.

Technology maturity

AI technology needed for a fully autonomous, humans-out-of-the-loop driving is not yet mature. The famous Moore's law is still predominant in the semiconductor sector.

It states that the number of transistors on a microchip doubles every two years, implying a significant increase in processing power. The increase in processing power generates more and more data resulting in a Big Data revolution. The exponential rate of growth of data catalyzes the development of new and powerful algorithms. The latter demand more processing power and so the *processing-data-algorithms* cycle keeps rotating rapidly (Figure 11.4).

Data, algorithms, and processing power are at the heart of the IT revolution, which has given birth to the AI and machine learning paradigm.

According to the technological version of the "law of accelerating returns,"[29] the rate of technological progress increases exponentially over time. In other words, the "returns" of the technological evolutionary process like speed, power, and cost-effectiveness increase exponentially over time. This happens because any growing technology gives birth to a new technology, which in turn grows and further develops the mother driving

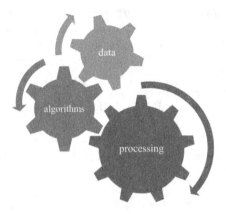

Figure 11.4 Data-algorithms-processing revolution.

technology. Meantime, other forms of supporting technologies get entangled and provide a positive feedback loop to the evolutionary process. The momentum picked up by the developing technology is forever accelerating. IoT promises to turn any ordinary everyday object into a computing and communicating device, thereby generating dynamic "Internet of Things" (IoT). Nanotechnology promises the production of new and resilient materials by manipulating materials at their atomic (nanometer range) level, and quantum computing promises to revolutionize the speed and volume of computations and will be game-changers giving an accelerated boost to the growth of AI and Machine Intelligence.

Self-driving vehicles are going to be just one application of this groundbreaking Machine Intelligence.

Cybersecurity

One of the unintended consequences of AVs is cybersecurity concerns. Car thieves once used mechanical devices to get into a car. Nowadays, they can use technology to hack the car system remotely. Future driverless cars will be entirely controlled by computer hardware and software. A malicious attacker could find and exploit security holes in any number of the millions of software components that dynamically make up the complex autonomous driving scenario. Attackers can take over AVs, orchestrate traffic jams, and set them on collision courses causing untold tragedies and loss to life and property. Following are some of the cybersecurity issues, AV designers and manufacturers need to address:

Sensor attacks

Sensor attacks can come in many forms. Distorting input signals from sensors through noise injection (jamming attack), generating fake sensor signals to present non-existing objects and phenomena (spoofing attack), shining strong light directly on a camera to impair its visibility (blinding attack), capturing pulses transmitted by the sensor, and resending them at different times and positions (replay attack) are some of the conceivable sensor attacks.[30]

Efficient hardware and software are yet to be designed to detect and counter such sensor threats.

Hardware attacks

Hardware acceleration with GPUs highly tailored for a specific task with their own memory and processor cores are the workhorses of the real-time computing machinery of AVs. Stealthily hidden security holes at the hardware level can give hackers ample chance towards an undesired manipulation of AVs.

Software attacks

Any modern software system is like an organic body composed of thousands of interacting components. The complexity of the system is such that it does not allow administrators to trace all possible loopholes. Automakers with links to IT enterprises are racing to address the security issues. For example, the E-Safety Vehicle Intrusion Protected Applications (EVITA)[31] is a noteworthy European automotive cybersecurity project. The EVITA Hardware Security Module (HSM) specification has become one of the major hardware standards in the automotive industry.

Infrastructure and network attacks

Driverless cars of the future will likely be networked to communicate with each other and the road infrastructure. V2V, V2I, and V2X are the three types of networks comprising Intelligent Transportation System (ITS) already under experimental operation in autonomous driving. Vehicle-to-vehicle communication technology (V2V) enables the exchange of information about position and speed from one vehicle to another. V2V networks uncover potential risks and dangers to driving and help reduce traffic congestion and accidents.

Vehicle-to-infrastructure communication (V2I) enables the exchange of information between vehicles and the traffic infrastructure. For example, smart traffic signals powered by V2I help AVs to adjust their speeds to smoothen their driving and reduce congestion. It also helps in smart self-parking. Vehicle-to-everything communication (V2X) encompasses both V2V and V2I technology. It gives power to AVs to communicate with the traffic system, including other vehicles and infrastructure. V2X can notify AVs and human drivers alike of road and weather conditions, accidents, and traffic congestion in the vicinity. Electronic Toll Collection (ETC) and self-parking in smart city parking lots operate on the V2X technology. The sheer size and complexity of the network components and their intra-communication protocols make it extremely difficult for security experts to make an exhaustive check of all possible security loopholes. Nonetheless, it is worth investing time, effort, and money checking the cybersecurity holes in such complex systems because attacks on the communication networks could maim the ITS creating chaos and incalculable harm to the robotic cars along with their occupants.

AVs IMPACT ON ECONOMY

The economic impact of self-driving vehicles will unfold itself in several stages. The benefits of AVs like safety, increase in people's mobility, and reduction in congestion, pollution, parking space, and cost will slowly transform society and auto industry. As more and more fleets of AVs are pressed in service, the revenues generated and the benefits reaped will far outnumber the unintended consequences AV technology may produce.

Although a fully matured level 5 AV technology is not yet available, level 3+ AVs are already being approved by governments for limited use. Industries such as manufacturing, farming, and construction are employing more and more self-driven trucks designed to accomplish specific, repetitive tasks in those fields. They are found to be safe, cost effective, safe, and indispensable in situations too risky for human involvement. Truck platooning which consists in the linking of several trucks in convoy on a highway is also on the rise. Using connectivity and autonomous driving technology, the trucks automatically maintain a set, close distance between each other. The truck at the head of the platoon acts as the leader with the vehicles behind following in unison its driving lead. Driving safety coupled with efficiency and lowering of fuel consumption and CO_2 emission is reportedly leading the truck platooning technology towards greater market share. Full-fletched platooning highway drive is expected to hit the market around 2023.[32]

In urban areas, robotaxis, automated school, and transit bus services are also on the rise. The accessibility, comfort, and ridesharing at a reduced price are attracting more and more customers to avail of the automated driving technology.

Currently, the price of a self-driving vehicle is exorbitantly high, but enterprises are pushing hard to make the technology economically viable. The LIDAR sensing technology, for instance, which measures distance using laser light to generate highly accurate 3D maps of the world around the AV, is considered by most in the self-driving car industry a key piece of technology required to safely deploy AVs. A single piece which was costing

$75,000 a while ago is now available at $7,500. Sensor technology and the rest of the hardware and software development costs will spirally decrease as more and more enterprises enter a fierce competition to get a share of the market which will put their AVs on public roads. It is only a matter of time before the current exorbitant price of AVs winds down to an affordable one. Image the computers of the past. They were mammoth structures occupying entire floors of university buildings. Individuals could not own them. The affordability and widespread use of personal computers today have proved to be beyond the wildest predictions of computer firm CEOs, entrepreneurs, and economists.

The auto industry is channeling billions of dollars into AV technology. Trillions of dollars are expected in return.[33] According to a cost estimate computed by Burns a tailor-designed, electrically driven, two-person, shared, AV will cost $0.20 per vehicle mile, compared to the $1.50 estimate of the cost of owning and operating a car. The gain is a whopping $1.30 per mile. Multiplying the $1.30 per mile savings by the 3 trillion miles Americans drive annually revealed AV mobility could reduce America's $4.5 trillion per year mobility bill by $3.9 trillion per year. The AV mobility solution could save a driver $5,625 a year, plus the value of all the time that not driving frees up. Converting the value of time into earnings, the annual savings could amount to $16,000 a year, or even higher.[34] According to Allied Research, the self-driving vehicle market is expected to grow from $54.23 billion in 2019 to $556.67 billion in 2026. In no time it will escalate to the order of trillion dollars.[35] These numbers support the rising trend of investment in the research and development of AV technology.

The autonomous driving revolution has begun, led not by the outstanding auto manufacturers, but by the IT companies. Some of these companies played a major role in amassing Big Data and producing cutting-edge machine learning algorithms for churning Big Data to arrive at profit-making business predictions and decisions. For them, boldly experimenting with autonomous driving using the newly developed generation and predictive processing of Big Data was the next logical step.

When large-scale experimenting showed signs of success, they began to invest heavily.

Traditional auto manufacturers are joining hands with IT companies to build AVs.

Academics, too, are not behind in the race for developing self-driving cars. They create simulation software and learning algorithms that clock millions of miles of safe driving on computer screens hidden behind the walls of the university research labs. For example, University of Michigan has taken a lead in creating a 32-acre, 10-million-dollar "Mcity" that provides a real-world space to test automated mobility systems. Mcity is a realistic scenario to evaluate how autonomous driving will shape urban planning.[36]

The universities involved in AV technology also produce engineering and data science graduates who are an asset to further boost the technology. University-developed self-driving platforms and prototypes along with their graduates well-trained and equipped to shape autonomous driving technology act as a bait for instant collaboration with industry. The auto manufacturing industry – IT companies – University Engineering Labs collaboration is just the right mixture to accelerate the progress of AVs to level 4 and 5 within a short time.

The self-driving car business is currently steeped in regulatory and ethical controversy, and the debate between the cons and pros is at a standstill.[37]

Exaggerated reports on fatal crashes involving AVs and statistical simulations showing that "fully autonomous vehicles would have to be driven hundreds of millions of miles and sometimes hundreds of billions of miles to demonstrate their reliability." have drawn excessive public frowning on AV technology, leaving an impression that the baby is thrown out with the bathtub. It is time that governments, policymakers, and entrepreneurs take a hard look at the benefits AV technology has the potential to deliver. The poignant question is: Does the public have to wait for years on end till it is proved that autonomous driving produces 0% accidents or do something using the available technology to address the sober fact that annually 1.35 million people die on the roads?

CONSOLIDATION WORKSHOP

1. Discuss the safety aspects of self-driving cars. What are the most likely challenges to the safety of self-driving cars?
2. Are currently available AVs safer than human driving? Will level 4, 5 AVs be better and safer than humans in driving? Discuss the philosophical and practical aspects of safety in the context of levels 4 and 5.
3. What is AV engineering? How does AI play a role in AV?
4. Often it is claimed that AVs will fail to recognize road signs and the like in mist, rain, and snow. Are humans smarter driving in bad weather conditions? Discuss keeping AI and NI combination in mind.
5. AV antagonist claim AVs will come across unforeseen conditions many times. What are some of these "unforeseen conditions" and what is the frequency and severity of their occurrence?
6. AV control systems are based on DL algorithms. What are the limits of these algorithms?
7. What security precautions should auto manufacturers take when designing AVs?
8. Discuss the safety, health, environmental, and economic benefits of AVs.

9. How long will it take for level 5 AVs to be a common place on public roads?
10. What are the chances of hackers gaining control of AVs and thwart the AI controls to cause accidents?
11. When an AV causes an accident who is to blame? Discuss the legal implications.
12. What are some of the new laws and regulations that need to be enforced before AVs get on public roads?

NOTES

1. Claudia Adriazola-Steil, Subha Ranjan Banerjee and Anna Bray Sharpin, Report---1.35 Million people killed every year in traffic crashes and counting. TheCityFix, January 9, 2019. https://thecityfix.com/blog/report-1-35-million-people-killed-every-year-in-traffic-crashes-and-counting-claudia-adriazola-steil-subha-ranjan-banerjee-anna-bray-sharpin/.
2. Burns, Lawrence, Shulga Christopher, *Autonomy: The Quest to Build the Driverless Car—And How It Will Reshape Our World*, Ecco; Reprint Edition, August 28, 2018.
3. https://newsroom.aaa.com/ 2020/03/self-driving-cars-stuck-in-neutral-on-the-road-to-acceptance/.
4. Dan Carney, How and why automakers work hard to camouflage their cars, Autoblog, November 7, 2014. https://www.autoblog.com/2014/11/07/how-and-why-automakers-work-hard-to-camouflage-their-cars/.
5. S. Chen, Y. Chen, S. Zhang and N. Zheng, "A Novel Integrated Simulation and Testing Platform for Self-Driving Cars with Hardware in the Loop," *IEEE Transactions on Intelligent Vehicles*, vol. 4, no. 3, pp. 425–436, September 2019, Doi: 10.1109/TIV.2019.2919470.
6. Rina Komatsu, Tad Gonsalves---Traffic Signs Automatic Recognition Using Convolution Neural Network, SDPS 2017, December 2017.
7. Junta Watanabe and Tad Gonsalves, In-vehicle camera images prediction by Generative Adversarial Network, *Proc. 6th International Conference on Computer Science & Information Technology*, CoSIT2019, February 23–24, 2019, Dubai, pp. 45–55.
8. https://www.synopsys.com/automotive/autonomous-driving-levels.html.
9. McGrath, Michael Eliot. *Autonomous Vehicles: Opportunities, Strategies and Disruptions---Updated and Expanded*; Second Edition. Kindle Edition. November 7, 2019.
10. https://www.asirt.org/safe-travel/road-safety-facts/.
11. https://www.viatech.com/en/ 2019/06/the-benefits-of-self-driving-cars/.
12. https://www.japantimes.co.jp/opinion/ 2019/06/15/editorials/preventing-elderly-driver-accidents/.
13. https://inrix.com/press-releases/ 2019-traffic-scorecard-us/.
14. Winston, Clifford, Karpilow, Quentin. Autonomous Vehicles. *The Road to Economic Growth?* Brookings Institution Press. Kindle Edition, June 30, 2020.

15. McKenzie, Hamish. *Insane Mode---How Elon Musk's Tesla Sparked an Electric Revolution to End the Age of Oil.* Dutton, November 27, 2018.
16. The Antiplanner, Transportation, January 30, 2017. (https://ti.org/antiplanner/?p=12818.
17. Thales, 7 benefits of autonomous cars, July 21, 2017. https://www.thalesgroup.com/en/markets/digital-identity-and-security/iot/magazine/7-benefits-autonomous-cars.
18. https://www.vtpi.org/avip.pdf.
19. S. Karnouskos, "Self-Driving Car Acceptance and the Role of Ethics," *IEEE Transactions on Engineering Management*, vol. 67, no. 2, pp. 252–265, May 2020, Doi: 10.1109/TEM.2018.2877307.
20. N. J. Goodall, "Can you program ethics into a self-driving car?" *IEEE Spectrum*, vol. 53, no. 6, pp. 28–58, June 2016, Doi: 10.1109/MSPEC.2016.7473149.
21. https://www.investopedia.com/articles/investing/090215/unintended-consequences-selfdriving-cars.asp.
22. Sibahle Malinga, *AI will Create More Jobs than it Destroys*, IT Web Business Technology Media Company, April 8, 2019. https://www.itweb.co.za/content/LPp6V7r4wQQqDKQz.
23. David Roe, *Why Artificial Intelligence Will Create More Jobs than it Destroys*, CMSWire, January 9, 2018. https://www.cmswire.com/digital-workplace/-why-artificial-intelligence-will-create-more-jobs-than-it-destroys/.
24. Zhang, X., Luo, H., Fan, X., Xiang, W., Sun, Y., Xiao, Q., ... Sun, J. (2017). Alignedreid: Surpassing human-level performance in person re-identification. *arXiv preprint arXiv:1711.08184.*
25. https://developer.nvidia.com/drive/drive-agx.
26. Insurance Institute for Highway Safety, Highway Loss Data Institute, Vol. 55, No. 3 July 22, 2020. https://www.iihs.org/api/datastoredocument/status-report/pdf/55/3.
27. Fallon, Michael, *Self-Driving Cars. Twenty-First Century Books* ™, August 1, 2018.
28. Eliot, Lance, AI self-driving cars divulgement: Practical advances. In *Artificial Intelligence And Machine Learning* (p. 1). LBE Press Publishing. Kindle Edition.
29. Ray Kurzweil, *The Age of Spiritual Machines: When Computers Exceed Human Intelligence*, Kindle Edition, Penguin Books, January 1, 2000.
30. Sjafrie, Hanky, *Introduction to Self-Driving Vehicle Technology* (Chapman & Hall/CRC Artificial Intelligence and Robotics Series). CRC Press. Kindle Edition.
31. Dr.-Ing. Olaf Henniger, Hervé Seudié, Robert Bosch, EVITA-Project.org---E-Safety Vehicle Intrusion Protected Applications, 7th escar Embedded Security in Cars Conference, November 24–25, 2009, Düsseldorf. https://www.evita-project.org/Publications/HS09.pdf.
32. https://www.acea.be/uploads/publications/Platooning_roadmap.pdf.
33. Alexander Hars, The auto industry is channeling billions into autonomous vehicle technology, August 31, 2018. http://www.driverless-future.com/.
34. See note 2.
35. https://www.globenewswire.com/news-release/ 2019/07/03/1877861/0/en/Global-Autonomous-Vehicle-Market-is-Expected-to-Reach-556-67-Billion-by-2026.html.

36. Angelo Rychel, 7 Universities that are pushing the boundaries of autonomous driving. Technology and Business, December 7, 2017. https://www.2025ad.com/7-universities-that-are-pushing-the-boundaries-of-autonomous-driving.
37. https://mobility.here.com/learn/smart-transportation/self-driving-car-levels-benefits-and-constraints.

Appendix A: Frameworks and libraries for ML

INTRODUCTION

Appendix A lists the top ten popular and oft-used open-source Machine Learning Frameworks & Libraries (MLFL). These MLFL are provided here based on the authors' understanding of them. There are potentially many more MLFL that readers would like to explore. This appendix provides a good starting point for eager practitioners to jumpstart their ML projects. Each framework/library has ample documentation on its website.

A.1 SCIKIT-LEARN

scikit-learn is an easy-to-use library interfaced with Python programming language. It is popularly used to conduct basic ML studies like supervised and unsupervised learning. It is built on packages like NumPy, SciPy, and matplotlib. The resulting libraries can be either used for interactive workbench applications or embedded into other software and reused.

https://scikit-learn.org/stable/user_guide.html

A.2 TENSORFLOW

TensorFlow, invented by Google, is an end-to-end open-source machine learning platform. It has industry-standard open-source libraries, tools, and community resources. It is both CPU and GPU compatible.

TensorFlow implements data flow graphs, where batches of data called "tensors" can be processed through a series of steps described by a graph. The movements of the data through the system are called "flows" – hence, the name TensorFlow. Graphs can be assembled with C++ or Python and can be processed on CPUs or GPUs.

It has tutorials for beginners, generative adversarial networks (GAN) and neural machine translation with attention for experts. Google, DeepMind, airbnb, CocaCola, GE Healthcare, and Intel are some of the companies using TensorFlow for research.

https://www.tensorflow.org/

A.3 KERAS

Keras is a simple machine learning framework for deep learning. It was developed with the objective of fast experimentation. It offers consistent and simple APIs which minimize user actions required for common use cases, and provides intuitive and actionable error messages. It also has extensive documentation and developer guides. Keras takes advantage of the full deployment capabilities of the TensorFlow platform. Keras models can be exported to JavaScript to run directly in the browser, to TF Lite to run on iOS, Android, and embedded devices. Keras is used by CERN, NASA, NIH, and many more scientific organizations around the world.

https://keras.io/

A.4 CAFFE

CAFFE, used by Facebook as a deep learning framework, has an excellent visual interface, modularity, and speed. It was developed by the Berkeley Vision and Learning Center (BVLC) and by community contributors. Models and optimization are defined by configuration without hard-coding such that users can switch between CPU and GPU.

https://caffe.berkeleyvision.org/tutorial/

A.5 PYTORCH

PyTorch, pioneered by Facebook, is a flexible and lightweight machine learning framework built for high-end efficiency. It runs on Python and has support for cloud environment. The deep learning research platform is a replacement for NumPy to use GPU acceleration.

https://pytorch.org/

A.6 THEANO

Theano is a Python library that lets the user define, optimize, and evaluate mathematical expressions, especially the ones with multidimensional arrays. Using Theano, it is possible to attain speeds rivaling handcrafted C implementations for problems involving large amounts of data. It supports the rapid development of efficient machine learning algorithms. With consistent and simple APIs, it minimizes the number of user actions required for common use cases and provides clear and actionable error messages. It also has extensive documentation and developer guides.

http://deeplearning.net/software/theano/

A.7 SPARK MLLIB

MLlib is Spark's machine learning library. Its goal is to make practical machine learning scalable and easy. It provides tools such as ML algorithms for machine learning; featurization for feature extraction and dimensionality reduction; and pipelines for constructing, evaluating, and tuning ML pipelines.

Spark ML provides a uniform set of high-level APIs that help users create and tune practical machine learning pipelines. Spark ML can run in clusters. In other words, it can handle large matrix multiplication by taking slices of the matrix and running that calculation on different servers.

http://spark.apache.org/docs/latest/ml-guide.html

A.8 THE MICROSOFT COGNITIVE TOOLKIT

The Microsoft Cognitive Toolkit (CNTK) is an open-source toolkit for distributed deep learning. It describes neural networks as a series of computational steps via a directed graph. CNTK can handle feed-forward DNNs, convolutional neural networks (CNNs), and recurrent neural networks (RNNs/LSTMs). CNTK also offers parallelization across multiple GPUs and servers.

https://docs.microsoft.com/en-us/cognitive-toolkit/

A.9 DEEPLEARNING4J

Deeplearning4j is a commercial and industry-focused Java-based deep learning framework. The platform supports CNNs, RNNs, recursive neural tensor network (RNTN), and long short-term memory (LSTM). It has GPU support and is widely used in industry for image recognition, fraud detection, text-mining, parts of speech tagging, and natural language processing.

https://deeplearning4j.org/

A.10 ONNX

ONNX (Open Neural Network Exchange) is an open standard for machine learning interoperability. Each computation dataflow graph in ONNX is structured as a list of nodes that form an acyclic graph. Nodes have inputs and outputs. Each node is a call to an operator. The graph also has metadata to help document its purpose, authorship, and so on.

https://onnx.ai/about.html

Appendix B: Datasets for ML and predictive analytics

INTRODUCTION

It is generally a tedious process to collect and preprocess data to conduct ML and predictive analytics. It is equally troublesome finding relevant datasets for data science and machine learning projects. Luckily, a large number of nonproprietary datasets are available for free download and use in the public domain. These are not just toy datasets aimed at data science and ML beginners to test their newbie programs, but rather large and real-life datasets. They cover virtually any domain under the sky. Appendix B reports the top 30 of the sites holding large collections of datasets. Beginners and experienced programmers can jumpstart their data science journey with these datasets.

1. Amazon Web Services – contain large public datasets which can be used for Big Data analytics on the cloud. Usage examples for all datasets are listed in the Registry of Open Data on AWS. The registry also contains a list of publications based on the analysis of datasets.
   https://aws.amazon.com/
   https://registry.opendata.aws/usage-examples/
2. bigml – holds a wealth of links pointing out to free and open datasets that can be used to build predictive models.
   https://blog.bigml.com/list-of-public-data-sources-fit-for-machine-learning/
3. Carnegie Mellon University (CMU) Libraries – CMU libraries provide high-quality datasets including the latest on COVID-19.
   https://guides.library.cmu.edu/machine-learning/datasets
4. Data.gov – publicly available datasets from federal, state, and local governments, including economic, geological, demographic, and many other types of data sources. This site also includes a list of other

Open Data Sites with similar publicly available data sources from various cities, states, and countries.

https://catalog.data.gov/dataset

5. Data.world – is geared for everyone – not just the "data people" – to get clear, accurate, fast answers to any business question. The cloud-native data catalog maps the user's data to the well-known business concepts, creating a unified body of knowledge anyone can understand and use.

https://data.world/

6. Deep learning – offers datasets for benchmarking deep learning algorithms. Datasets categories include natural images, synthetic datasets, faces, text, speech, recommendation systems, and miscellaneous.

http://deeplearning.net/datasets/

7. EliteDataScience – offers a curated list of free datasets for data science and machine learning, organized by their use case.

https://elitedatascience.com/datasets

8. FAOStat – data portal for the Food and Agriculture Organization for the United Nations. The site provides free access to food and agriculture data for over 245 countries and territories, and covers all FAO regional groupings from 1961 to the most recent year available.

http://www.fao.org/faostat/en/#home

9. Google Dataset Search – Google Dataset Search engine helps the user find datasets wherever they are hosted, whether it's a publisher's site, a digital library, or an author's web page. It is a high-end dataset finder, having access to over 25 million datasets.

10. Humans in the Loop – upon request, creates custom ML datasets by gathering and curating bias-free data as close as possible to the user's request.

https://humansintheloop.org/dataset-collection/

11. IEEE ComSoc – IEEE Communications Society provides datasets and competitions on Machine Learning for Communications Emerging Technologies.

https://mlc.committees.comsoc.org/datasets/

12. Kaggle – Kaggle provides a vast number of datasets, sufficient for the enthusiast to the expert. The site allows users to find and publish datasets, and explore and build models in a web-based data science environment. Kaggle holds regular competitions to solve large-scale data science challenges.

https://www.kaggle.com/

13. KDnuggets – an online platform for business analytics, Big Data, data mining, and data science. The platform covers analytics and data mining, including news, software, jobs, meetings, courses, data, education, and webinars.

https://www.kdnuggets.com/

14. Lionbridge AI – has assembled a wealth of resources for machine learning and natural language processing activities. The website explains the use of a wide number of open datasets ranging from the general to the highly specific, such as financial news or Amazon product datasets.
    https://lionbridge.ai/datasets/the-50-best-free-datasets-for-machine-learning/

15. London Datastore – is a free and open data-sharing portal where anyone can access data relating to London. It is open to citizens, business persons, researchers, developers, and so on. The site provides over 700 datasets to help the user understand the city of London and develop solutions to London's problems. The portal was a winner of the 2015 ODI Open Data Publisher Award.
    https://data.london.gov.uk/

16. Machine Learning Mastery – offers ten widely used standard datasets for practicing Applied Machine Learning.
    https://machinelearningmastery.com/standard-machine-learning-datasets/

17. Machine-Learning-Tokyo – displays several public datasets for ML and predictive analysis.
    https://github.com/Machine-Learning-Tokyo/public_datasets

18. Microsoft Datasets – a collection of free datasets from Microsoft Research to advance state-of-the-art research in areas such as natural language processing, computer vision, and domain-specific sciences. Datasets can be downloaded directly to a cloud-based Data Science Virtual Machine for a seamless development experience.
    https://msropendata.com/

19. Million Song Dataset – freely available collection of audio features and metadata for a million contemporary popular music tracks.
    http://millionsongdataset.com/

20. Movielens – Real movie ratings data contains ratings on 1600+ movies by 1000 users. It helps find the movies of the user's liking, rates the movies to create a custom taste profile, and finally recommends other movies to the user based on the user's taste profile.
    www.movielens.org

21. ODSC – The Open Data Science site lists 25 Excellent Machine Learning Open Datasets on NLP, sentiment analysis, reviews, government data, finance and economics, and health and image data.
    https://medium.com/@ODSC/25-excellent-machine-learning-open-datasets-940ca2124dfc

22. Open Graph Benchmark – OGB provides a diverse set of challenging and realistic benchmark datasets that are of varying sizes and cover a variety of graph machine learning tasks, including prediction of node, link, and graph properties.
    https://ogb.stanford.edu/

23. OpenML – boasts 21,154 datasets, 217,369 tasks for scientific analysis, 15,926 flows, and 10146188 runs. There is also an option to upload and explore all results online.
    https://www.openml.org/

24. SNAP – Stanford Large Network Dataset Collection (SNAP) contains datasets for social media, Wikipedia networks, Amazon networks, online communities, online reviews, and so on.
    https://snap.stanford.edu/data/

25. UCI Machine Learning Repository – a repository containing 559 datasets for data mining and machine learning, maintained by UC Irvine Center for Machine Learning and Intelligent Systems
    http://archive.ics.uci.edu/ml/index.php

26. US Census Bureau – data portal of the US Census Bureau.
    https://www.census.gov/data.html

27. Visual Data – computer vision datasets can be searched by category; it allows flexible searchable queries.
    https://www.visualdata.io/discovery

28. Where to Find the Best Machine Learning Datasets – contains several datasets on a wide range of topics ranging from COVID-19 stats to Harry Potter spells. The site also shows how to search for ML-specific datasets.
    https://serokell.io/blog/best-machine-learning-datasets

29. Yelp – The Yelp dataset offers reviews and user data for use in personal, educational, and academic purposes. The data is available as JSON files which can be used to teach students about databases, to learn NLP, or for sample production data to make mobile apps.
    https://www.yelp.com/dataset

30. 70+ Machine Learning Datasets and Project Ideas – offers more than 70 machine learning datasets that can be used to build data science projects.
    https://data-flair.training/blogs/machine-learning-datasets

Appendix C: AI and BO research areas

- Where else can data be applied? Data is usually applied linearly. Weather data for weather predictions. Orthogonal application of data.
- What are the risks (GRC) associated with data application? Privacy, security.
- Nature of Data. Transactional, macro, alternative
- Data to Decisions pyramid
- Advances in machine learning, deep learning
- Cognitive computing, intelligent agents
- Chatbots embedded with intelligence
- AI strategy for business and industry
- AI applications in industry, business, healthcare, and education and training
- AI in government and legal practice
- Entertainment in the age of AI – movies, sports, theater, concerts
- AI for enhancing information security and privacy
- Work style, ethics, protocols in the age of AI
- Trust, resilience, privacy, and security issues in AI applications
- Testing and validation of AI and ML applications
- Risks, limitations, and challenges of AI and ML
- Legal, regulatory, ethical aspects of AI (liability, etc.)
- AI: promise vs practice
- Societal implication of the rise of AI
- Human-machine coexistence and collaboration
- Intelligent, autonomous robots, automated cars, and drones
- Industry 4.0
- Smart cities to smart society
- AI and IoT – the server domain

- Supply chains
- Case studies, experience reports, position papers, and visionary perspectives
- Overview of AI activities specific to a region/country
- The future of AI

Index

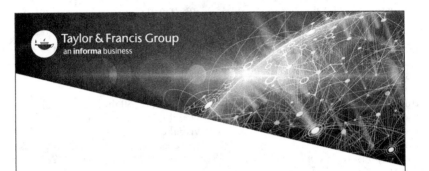

Printed in the United States
by Baker & Taylor Publisher Services